COMPUTER SECURITY IN THE 21ST CENTURY

COMPUTER SECURITY IN THE 21ST CENTURY

Edited by
D. T. LEE
Academia Sinica, Taiwan

S. P. SHIEH
National Chiao Tung University, Taiwan

J. D. TYGAR
UC Berkeley

Editors:

D.T. Lee
Academia Sinica, Taiwan

S.P. Shieh
National Chiao Tung University, Taiwan

J.D. Tygar
UC Berkeley

Computer Security in the 21st Century

Library of Congress Cataloging-in-Publication Data

Computer security in the 21st century / edited by D.T. Lee, S.P. Shieh, J.D. Tygar.
p. cm.
Includes bibliographical references and index.

1. Computer security. I. Title: Computer security in the twenty-first century. II. Lee, D. T. III. Shieh, S. P., 1960- IV. Tygar, J. D.

QA76.9A25C648 2005
005.8--dc22

2005042554

ISBN 978-1-4419-3679-0 e-ISBN 978-0-387-24006-0

Printed in the United States of America.

9 8 7 6 5 4 3 2 1

springeronline.com

Contents

List of Figures

Chapter 1

INTRODUCTION

D. T. Lee
Academia Sinica, Taiwan

S. P. Shieh
National Chiao Tung University, Taiwan

J. D. Tygar
UC Berkeley

Computer security has moved to the forefront of public concern in the new millennium. Hardly a day passes where newspaper headlines do not scream out worries about "phishing" , "identity theft" , "browser exploits" , "computer worms" , "computer viruses" , "online privacy" , and related concerns. The major vendor of computer operating systems has announced that computer security is now its top priority. Governments around the world, including most major governments in North America, Europe, and East Asia continue to worry about "cyber-terrorism" and "cyber-war" as active concerns.

It was in this charged environment that we decided to hold a workshop in December 2003 on emerging technologies for computer security. The workshop was held in Taipei in conjunction with several other conferences (notably Asiacrypt) and featured leading researchers from the Asia-Pacific region and the United States. What followed was three days of exchange of ideas that led to a number of significant developments. This book attempts to share some of the research trends that were reflected in the best papers published at the conference.

The first section deals with the classical issue of cryptographic protocols. How can we build secure systems that need to exchange private data, while guarding against eavesdroppers who listen in on attacks? Dieter Gollmann examines five case studies that show challenges in cryptographic protocol design and argues for a new framework for viewing the problem. Yaping Li, J. D.

Tygar, and Joseph Hellerstein show how private matching can be used to exchange database information while still protecting the privacy of individuals. Jonathan Millen brings formal analysis to bear, showing that current techniques of analyzing protocols still fail to protect against a number of problems. And Tzong-Chen Wu and Yen-Ching Lin argue for a a new key agreement method based on self-certification .

We next turn our attention to networking, and examine the rapidly expanding fields of peer-to-peer networking and ad hoc networking. These clearly introduce a number of new security challenges, and are especially relevant in light of recent studies suggesting the peer-to-peer networking now comprises the majority of networking over the Internet. Nitesh Saxena, Gene Tsudik, and Jeong Hyung Yi present a new system, Bouncer , that provides arguably the most fundamental element of peer-to-peer security: secure admissions control. They also discuss its actual implementation in several real peer-to-peer networks. And Shih-I Huang, Shiuhpyng Shieh, and S. Y. Wu present key distribution systems for an important emerging type of ad hoc network : wireless sensor networks .

A fundamental change in thinking about security has been the change of emphasis from building impenetrable systems to building systems that rapidly respond when attacks commence. Michael Howard, Jon Pincus, and Jeannette M. Wing report on work at Microsoft that proposes a completely new way of thinking about the vulnerability of systems: "relative attack surfaces ". Pei-Te Chen, Benjamin Tseng, and Chi-Sung Laih give a new may of modeling intrusion detection systems . Fu-Yuan Lee, Shiuhpyng Shieh, Jui-Ting Shieh, and Sheng-Hsuan Wang propose a new type of system for actively responding to distributed denial of service attacks; and Chang-Hsien Tsai, Shih-Hung Liu, Shuen-Wen Huang, Shih-Kun Huang, and Deron Liang discuss their BEAGLE system that allows security faults to be reproduced for debugging purposes.

Finally we turn our attention to perhaps the hottest single topic in the set of emerging security concerns: protecting multimedia content. Yao-Wen Huang and D. T. Lee discuss issues in Web Application Security. Robert H. Deng, Yongdong Wu, and Di Ma discuss their work in securing a new standard for photographic images, JPEG2000 . And Chin-Chen Chang, Tzu-Chuen Lu, and Yi-Long Liu discuss a new method of "watermarking" information in documents: a secret information hiding scheme.

Together, these works present an agenda of important security topics for computer security in the new century.

Acknowledgments

Support for this book came from Academia Sinica, National Chiao Tung University, and the University of California, Berkeley. D. T. Lee received additional support from National Science Council, and Science and Technology Advisory Group of Executive Yuan, Taiwan. Shiuhpyng Shieh received additional support from National Science Council, Ministry of Education, Taiwan. and Industrial Technology Research Institute. J. D. Tygar received additional support from the US National Science Foundation and the US Postal Service. The opinions in this book are those of the authors and do not necessarily reflect the opinions of the funding sponsors or any government organization.

I

SECURITY PROTOCOL DESIGN

Chapter 2

CHALLENGES IN PROTOCOL DESIGN AND ANALYSIS

Dieter Gollmann
TU Hamburg-Harburg
Germany
diego@tu-harburg.de

Abstract The clarification of protocol goals and of the assumptions made about the environment protocols are intended for is an important but sometimes underestimated step in protocol design and analysis. Implicit assumptions about the environment can profoundly influence our understanding of security and may mislead us when faced with new challenges. Five case studies will support these claims. Research on novel security properties and on the influence of assumptions about the environment are proposed as major challenges in protocol design and analysis.

Keywords: Protocol analysis, authentication, key establishment.

1. Introduction

Since Needham and Schroeder published their paper on authentication [Needham and Schroeder, 1978], the design and analysis of security protocols has been an active field of research. In Needham's words, this paper had the lasting effect of making research on three-line protocols socially acceptable. The BAN logic of authentication [Burrows et al., 1990] provided further impetus for research on the formal analysis of security protocols. Considerable progress has been made in developing increasingly powerful analysis methods, see e.g. [Ryan et al., 2001]. New attacks were found against protocols that had been in the public domain for years, and had even been the subject of prior formal analysis [Lowe, 1996, Lowe, 1995]. Given the origins of this research area, it is no surprise that many efforts were directed towards the aforementioned three-line protocols, and in particular towards authentication protocols. The discovery of new attacks was seen as evidence that the design of security protocols is "difficult and error prone" and that tool support is necessary as problems are too complex for analysis by hand.

We will argue that the perceived difficulties in protocol design are not so much due to the inherent complexities of the attacks that have been found, but to ambiguous protocol goals and changing assumptions about the environment a protocol is fielded in. For example, the report on the BAN logic of authentication explicitly avoids giving a definition of authentication and advertises the logic as a means of differentiating between goals "authentication" protocols might achieve. Authentication is quite an overloaded notion so misunderstandings can occur easily. Clarification of goals is thus a crucial step in protocol analysis.

The specification of novel security goals and research into the interrelations between protocol goals and the environment a protocol is intended for are therefore major challenges in protocol design and analysis. We will illustrate these points with five case studies. Section 4.1 revisits Lowe's attack against the Needham-Schroeder protocol and explains this attack as a result of changing assumptions about the intended environment and about the meaning of authentication. Section 4.2 shows that it is possible to establish secure connections without mutual entity authentication. Section 4.3 examines security issues of location management in Mobile IPv6 . Section 4.4 reflects on data integrity in sensor networks and shows that even generally accepted facts in security may be rooted in implicit assumptions about the environment. Section 4.5 refers to a study of key insertion attacks done more than a decade ago to point out a direction in protocol analysis where we can face truly complex attacks.

These case studies are preceded by preliminary discussions about the purpose of protocol analysis and its repercussions on analysis methods, and about fundamental changes in assumptions about the environments security protocols are being designed for.

2. Purpose of Analysis

In protocol analysis, we are given a communications environment together with a description of the adversary (sometimes implicitly), a set of security goals (sometimes expressed in anthropomorphic metaphors), and a protocol, and examine whether the protocol does meet its desired goals. Two distinct motivations can drive the development of methods and tools for protocol analysis.

- Analysis of protocols that address well established security requirements and use established security primitives.

- Analysis of protocols that address novel requirements.

In the first case, security goals, assumptions about the environment, and standard cryptographic primitives can be integral parts of the methodology. Methods of this kind are useful, for example, when dealing with system architectures that expose security to application writers. Systems where security is the responsibility of the application layer tend to fall into this category. Typically, developers are not security experts but are asked to integrate standard security mechanisms that meet standard goals into their designs. Changes to any of these aspects would require some redesign of the methodology, but this could be done "out-of-band" before a revised tool is handed back to the developers.

In the second case, we need *agile* methodologies where it is easy to define specific adversaries and to express novel security requirements. Mostly, new protocols are designed because new requirements are emerging so that traditional security assumptions have to be adjusted. The analysis of protocols meeting novel security requirements is the main focus of this paper.

3. The Environment

Any model for analysing security protocols will in some way or other capture assumptions about the environment the protocol is deployed in, including aspects of the communications network, the actions an adversary can take, and the behaviour of regular protocol participants. When interpreting the results of a protocol analysis, it is evidently important to understand which environment had been modelled.

3.1 Dolev-Yao

Often, protocol analysis tries to assume as little as possible about the communications systems and gives all messages to the adversary for delivery. Many authors refer to the Dolev-Yao model [Dolev and Yao, 1983] when taking this approach. This model makes two independent assumptions:

- Cryptography is "perfect". The adversary does not try to exploit any weakness in the underlying cryptographic algorithms but only algebraic properties of cryptographic operators and interactions between protocol messages.

- The adversary can observe and manipulate all messages exchanged in a protocol run and can itself start protocol runs.

The second assumption had already been stated by Needham and Schroeder [Needham and Schroeder, 1978]:

> We assume that the intruder can interpose a computer in all communication paths, and thus can alter or copy parts of messages, replay messages, or emit

> false material. While this may seem an extreme view, it is the only safe one when designing authentication protocols.

The Dolev-Yao model makes no explicit provision for describing different communications environments (other than by defining the derivation rules the adversary can apply) and the adversary is limited to exploiting specific algebraic properties of cryptographic operators.

Analysing protocols in a setting as general as possible is not necessarily a route to higher security. Protocols may build on particular features of their intended environment and should in the first place be analyzed in a model that is faithful to this environment. Showing that a protocol does not meet is goal in a more general setting is useful side-information but should not be automatically classified as an attack.

Thus, for any approach to protocol analysis we should query how different communications environments and adversaries can be modelled. If an approach works for a fixed environment only, we should be aware of its limitations. Fo illustration, we contrast the mechanisms and the assumptions underpinning security in closed and open environments.

3.2 Closed environments

Security research originated in places like research laboratories or university departments. In such closed organisations, users have identities (company ID, student ID), can be physically located, and are subject to the authority of other entities in the organisation (managers, heads of department). Assumptions specific to closed environments underpin many familiar approaches to security. In traditional computer security,

- security policies refer to user identities; access control consists of authentication (checking who you are) and authorisation (checking whether you have the necessary access rights) [Lampson et al., 1992].
- Access control defends against attacks by outsiders; principals are "honest" as stated by Needham [Needham, 2000]: If they [principals] were people they were honest people; if they were programs they were correct programs.
- Auditing is used to detect attacks by insiders: "If you break the rules we can get hold of you".

The anthropomorphic metaphor that principals are "honest" should not be interpreted as a general expectation that people within the organisation are honest. The metaphor just indicates that security mechanisms do not address threats from insiders. Obviously, when a verification method such as the BAN logic makes this assumption, any conclusion drawn is only valid in environments where the assumption holds.

3.3 Open environments

For our purpose, open environments are characterized by the absence of strict lines of authority. In the extreme case, parties may join and leave the system on their own terms. Some ad-hoc networks fit this description [Mäki and Aura, 2002]. As another example, consider parties joining together in a virtual organisation where there is some agreement between partners but no entity has real authority over the others.

Our emphasis is on organisational structures. In this respect, our concerns differ from those in open systems security as commonly understood in network security. There, we are concerned with closed environments connected by open networks. For security, the move to open environments has a number of implications.

- User identities may be of little value. Names are useful locally [Ellison et al., 1999] but in an open environment we may deal with users not previously known, whose name (identity) does not appear in any security policy, and who may be outside the reach of any authority we are able to invoke.

- Security policies use attributes other than user identity. Java security [Gong, 1999] and .NET security [La Macchia et al., 2002] have been moving to code-based (evidence based) access control for some time.

- There need not be a central authority for setting policies, e.g. in ad-hoc networks or in peer-to-peer networks.

- There need not be a central entity making access control decisions.

- There is no boundary between inside and outside. The enemy is within by default. Principals need not be honest.

If security policies no longer refer to user identities and if authentication checks who you are, we have access control without authentication. Conversely, if authorisation checks that your identity appears in an access control list and if security policies are completely encoded in certificates, we have access control without authorisation. Thus, even the language we use to discuss security is geared to closed environments and starts to fail us when we move to new settings.

A general challenge today is to expose closed system assumptions that are inappropriate in open environments, both in the security mechanisms we are familiar with and in the (formal) analysis methods at our disposal.

4. Case Studies

We discuss four security protocols and their analysis, focusing on the specification of security goals and on the underlying communications environment.

4.1 Lowe's attack

Protocol analysis using CSP and the model checker FDR came to prominence with Lowe's analysis of the Needham-Schroeder public key protocol [Lowe, 1996, Lowe, 1995]. The fact that his analysis found a previously unreported attack is often quoted as evidence for the benefits of using formal methods when analysing security protocols.

The Needham-Schroeder public key protocol was intended for establishing secure connections. When starting a session, parties A and B with public keys K_a and K_b, and in sole possession of the respective private keys, establish shared secrets. A secure connection for exchanging messages during the session is then created, for example, by deriving a session key from the shared secrets. The three messages at the core of the protocol are [Needham and Schroeder, 1978]:

1. $A \to B\colon eK_b(N_a, A)$
2. $B \to A\colon eK_a(N_a, N_b)$
3. $A \to B\colon eK_b(N_b)$

Only B can decrypt the first message and obtain N_a. Only A can decrypt the second message, obtaining N_b and a confirmation for its challenge N_a. In the third message B receives a confirmation for its challenge N_b. A session key can now be derived from the two nonces N_a and N_b. A formal proof in the BAN logic of authentication confirms that the protocol securely establishes shared secrets [Burrows et al., 1990]. The logic explicitly considers attacks by outsiders only.

In Lowe's analysis, the protocol goals are given as *correspondence properties*:

- The initiator should only commit if the intended responder had replied to its challenge.
- The responder should only commit to a protocol run with the initiator that had sent the challenge.

In the attack principal A starts a protocol run with an "evil" party E that manages to fool B and, depending on your point of view, also A:

1. $A \rightarrow E$: $eK_e(N_a, A)$
2. $E \rightarrow B$: $eK_b(N_a, A)$
3. $B \rightarrow E$: $eK_a(N_a, N_b)$
4. $E \rightarrow A$: $eK_a(N_a, N_b)$
5. $A \rightarrow E$: $eK_e(N_b)$
6. $E \rightarrow B$: $eK_b(N_b)$

After the last step, B commits to a protocol run with A although A had embarked on a protocol run with E and is unaware of B's involvement. B is deceived as E has impersonated A in a protocol run with B. Lowe also reasoned that A successfully authenticates E because it receives a valid reply to its initial challenge in step 4.

In a trace-based analysis , expressing security properties as correspondence conditions on traces comes natural. Often protocol specifications are amended with auxiliary *begin* and *end* events for this purpose. We could try to define authentication generically as a relation between two sets of events R and T. "T authenticates R" [Schneider, 1996], or "R precedes T" [Schneider, 1998], when every occurrence of an event from T in the trace of a protocol run must be preceded by the occurrence of an event from R.

We would get a range of formal properties reflecting different informal security requirements but we cannot take for granted that the goals of any given so-called authentication protocol will be captured by our property of choice. Not every correspondence property need relate to an established notion of authentication either and reference to auxiliary events introduces another degree of arbitrariness.

When correspondence properties were introduced [Bird et al., 1992], their purpose was stated as:

> Note that the requirement is that the *exchange* be authenticated, and not the parties themselves.

Today, the term entity authentication is used in this meaning:

DEFINITION 2.1 *Entity authentication is the process whereby one party is assured of the identity of a second party involved in a protocol, and that the second has actually participated [Menezes et al., 1997].*

However, this was not always the case. The Needham-Schroeder authentication protocols were designed to establish secure connections. In this regard, Lowe's analysis misses an attack on A. At the end of the attack above, E shares a key with A and B while A erroneously believes to share a key with E only and B believes to share a key with A.

Lowe's analysis also highlights the distinction between the traditional scenario where honest principals seek protection from an outsider interfering with their communications and settings where insider attacks are a concern. This change in assumption rather than the use of a model checker led to the discovery of the attack against the Needham-Schroeder public key protocol, an attack that is launched by a dishonest principal. This issue did not arise in the work of Dolev and Yao, who analyze the secrecy of data items sent in a protocol message. The Dolev-Yao model is thus silent on the nature of the adversary, and this aspect should be clarified in any reference to it.

4.2 Secure connections without entity authentication

The previous section raises the question how entity authentication (Def. 2.1) is related to establishing a secure connection. It may seem that entity authentication is a prerequisite for, say, establishing a shared session key . If a party cannot tell for certain whom it has executed a protocol run with, how can it tell that it is not connected with an adversary?

However, this view is wrong as shown in [Mitchell and Pagliusi, 2003]. Their counterexample is set in an environment typical for mobile telecommunications systems like GSM (Global System for Mobile Communications). There is a mobile device, a home server, and a visited server. Mobile and home server share a secret key K_i.

In GSM, when the mobile tries to register with the visited server, the server contacts the home server and receives a triplet (N, R, K_c), where N is a random challenge, R the expected response, and K_c a session key for optionally encrypting the channel between visited server and mobile. Both R and K_c are derived from N and the key K_i. The visited server then sends the challenge N to the mobile, which computes the response R and key K_c from N and K_i, and returns its response. Authentication succeeds if returned response and expected response match.

DEFINITION 2.2 *Key authentication is the property whereby one party is assured that no other party aside from a specifically identified second party may gain access to a particular secret key [Menezes et al., 1997].*

The GSM protocol provides unilateral authentication only. The visited network authenticates the mobile, but not vice versa. The protocol cannot preclude man-in-the-middle attacks where the adversary inserts itself between mobile and visited server. Mutual key authentication is often suggested as the remedy.

In GSM, mutual key authentication would not help as there is no integrity protection for the subsequent voice traffic. Conventional cryptographic integrity protection of voice traffic is no option because the error rate on the wireless channel is too high. Transmission errors would invalidate too many data frames. The optional encryption would stop the attack, but in GSM the adversary can tell the mobile to switch off encryption. Moreover, the protocol does not provide *key freshness*. An adversary who has obtained a triplet

(N, R, K_c) can mount a replay attack when the mobile makes a connection requests.

Mitchell and Pagliusi constructed the following protocol specifically to show that man-in-the-middle and replay attacks can be stopped without resorting to mutual authentication [Mitchell and Pagliusi, 2003]. The environment is as in the GSM example, with the addition that mobile and home server have synchronized clocks. The home server provides the visited server with a 5-tuple (N, R, K_c, K_a, T) where

- challenge N and response R are as in the GSM protocol,
- T is a timestamp,
- K_c, K_a are keys derived from N, T, and the shared secret key K_i.

The visited server sends N and T to the mobile. The mobile checks whether T is within an acceptable time window and aborts the protocol if T is too old. This check prevents replay attacks. If T is accepted, the mobile computes R, K_c, and K_a from N, T, and K_i. As before, the visited server matches expected and received response to authenticate the mobile.

The key K_c is again an encryption key for the data channel between mobile and visited server. The key K_a is used to protect the integrity of the signalling channel between mobile and visited server. Thus, the adversary cannot modify or forge signalling requests, e.g. asking the mobile to switch off encryption. When encryption is switched on, the key K_c provides the mobile with a connection the adversary cannot eavesdrop on. Still, the mobile gets no evidence whatsoever about the party it is running the protocol with. The mobile gets the keys to build a secure connection without conducting entity authentication.

4.3 Location management in Mobile IPv6

Network addresses can serve two general aims. An address can uniquely identify a node in the network, or encode the location of a node in the network topology. As long as nodes remain in fixed positions it may not be necessary to distinguish between identity and location. When nodes are mobile, identity and location are clearly separate concepts. In Mobile IPv6, the 128-bit IP addresses consist of a 64-bit interface identifier and a 64-bit routing prefix (location information).

Mobile nodes can lie about their identity or their location. An adversary can fraudulently present its own location as the victim's current address to hijack a connection or give the victim's address as its own location to mount a flooding attack . Thus, we need protocols for verifying that a node with a claimed identity is in its claimed location.

We sketch a binding update protocol for mobile IPv6 [Aura et al., 2002], assuming the following about the communications infrastructure. Each mobile node has an address in its home network. Messages sent to this home address are routed to the mobile node via a secure IP tunnel . For better efficiency, a mobile node can inform a correspondent node about its current location by performing a *binding update* .

The adversary can eavesdrop on traffic on the wireless links used by the mobile node, make arbitrary claims about its current location, and use arbitrary identities when sending messages to the correspondent. We assume that the channel between home network and correspondent is reasonably secure, e.g. because it uses the wired Internet which could be secured by other means. For detailed justifications the reader is referred to [Aura et al., 2002]. We have thus three channels, with different security characteristics.

- Mobile $\leftrightarrow$ home: the mobile node and its home network have a pre-arranged security association, which they can use to create a secure IP tunnel to transfer messages.
- Correspondent $\leftrightarrow$ home: uses the wired Internet.
- Mobile $\leftrightarrow$ correspondent: unprotected wireless channel.

In the protocol in Fig. 2.1, the mobile node first sends binding update requests (BU) to the correspondent, via the home network and directly over the radio channel (steps 1a and 1b). The correspondent replies to both requests independently, sending a key K_0 to the mobile node via the mobile's home address and a second key K_1 directly to the claimed current location (steps 2a and 2b). The mobile node uses both keys to compute a message authentication code for the binding update (step 3).

This protocol does not rely on the total secrecy of cryptographic keys. In the threat model chosen, it is admissible to send the keys K_0 and K_1 in the clear on the channels from the correspondent. Technically, these keys can be interpreted as challenges (nonces) that bind identity to location through the

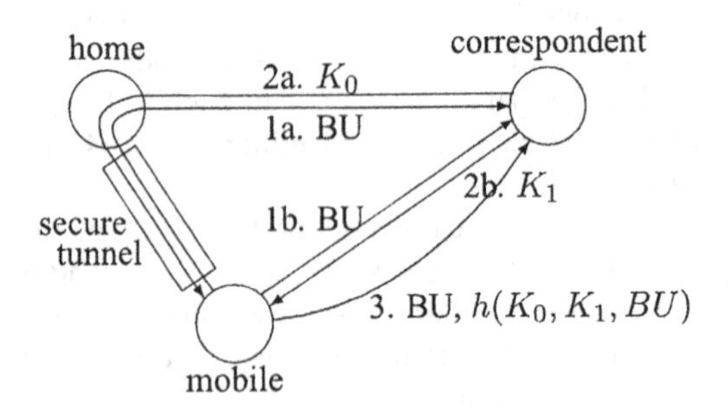

Figure 2.1. Binding updates in Mobile IPv6

hash $h(K_0, K_1, BU)$. The fluid border between keys and nonces was already evident in Section 4.1.

As discussed in [Gollmann, 2003b], in communications security the term authentication habitually refers to the corroboration of a link between an identity of some kind and an aspect of the communications model, like a message or a session. In this respect, binding update protocols provide *location authentication* .

4.4 Data integrity in sensor networks

Consider a network with the following characteristics [Vogt, 2003]. Nodes can communicate with their direct neighbours. Nodes only have information about nodes in their vicinity and no means for authenticating arbitrary nodes in the network.

In cryptographic terms, there is nothing like a public key infrastructure. Nodes share secret keys with nodes that are one or two hops away. Message authentication codes protect the integrity of messages transmitted between nodes that have shared keys.

Nodes can inject new messages into the network and forward messages they receive. A discussion of suitable routing algorithms for such a network is outside the scope of this paper. We assume such an algorithm exists. A discussion of potential applications of such networks is also outside the scope of this paper. In such a network we want to achieve *data integrity*. Forwarded messages cannot be manipulated or inserted[1]:

DEFINITION 2.3 *Data integrity is the property whereby data has not been altered in an unauthorized manner since the time it was created, transmitted, or stored by an authorized source [Menezes et al., 1997].*

Defence against the creation of messages with bad content is a separate issue that is not being addressed here.

In the network given, the creator of a message cannot vouch for its integrity as nodes further away would not share a key with the originator. The Canvas protocol thus uses interwoven authentication paths for data integrity [Vogt, 2003]. We slightly modify the original presentation and include the identities of nodes on the routing path with messages. We only describe the forwarding of messages. The injection step in [Vogt, 2003] is constructed along similar lines.

Let K_{xy} denote a symmetric key shared by nodes X and Y and let A, B, C, D, denote nodes in the network. A message m is forwarded from B to C as follows (Fig. 2.2):

$$B \to C : m, A, B, D, h(K_{ac}, m), h(K_{bc}, m), h(K_{bd}, m)$$

Node A had forwarded the message to B, nominated C as the next node, and included the authenticator $h(K_{ac}, m)$. In turn, B nominates D as the next node and constructs authenticators $h(K_{bc}, m)$ and $h(K_{bd}, m)$. The recipient C checks the two authenticators $h(K_{ac}, m)$ and $h(K_{bc}, m)$, and discards m if authentication fails.

Obviously, if A and B collude they can modify m without being detected by C. However, it can be shown that the protocol achieves its goal if no two adversarial nodes are direct neighbours [Vogt, 2003]. This observation contradicts a view widely held in communications security that *data integrity* and *data origin authentication* are equivalent properties, see e.g. [Menezes et al., 1997, page 359].

DEFINITION 2.4 *Data origin authentication (message authentication) is a type of authentication whereby a party is corroborated as the source of specified data created at some time in the past [Menezes et al., 1997].*

By definition, data origin authentication includes data integrity. In a communications system where the sender's identity (address) is an integral part of a message, a message with a forged sender address should not be accepted as genuine. To check the integrity of a message we would also have to verify its origin. Moreover, if messages pass through a completely insecure network, we can only rely on evidence provided by the sender to verify that a message has not been altered in transit. For both reasons data integrity includes data origin authentication.

However, in an open environment where the identities of other parties are unknown the sender's identity may not be an integral part of a message. Furthermore, if we do not assume that the network is completely insecure, we might accept that a message is received exactly as it was created if a sufficient number of independent witnesses can vouch for this fact. We may have to rely in turn on other witnesses to confirm witness statements. In such a setting we can have data integrity without data origin authentication.

There is a final twist to this case study. We can find an attack against the protocol by adjusting assumptions about the adversary. Adversarial nodes still cannot be direct neighbours in the network but they may agree a-priori on a strategy for modifying messages and know their respective routing strategies.

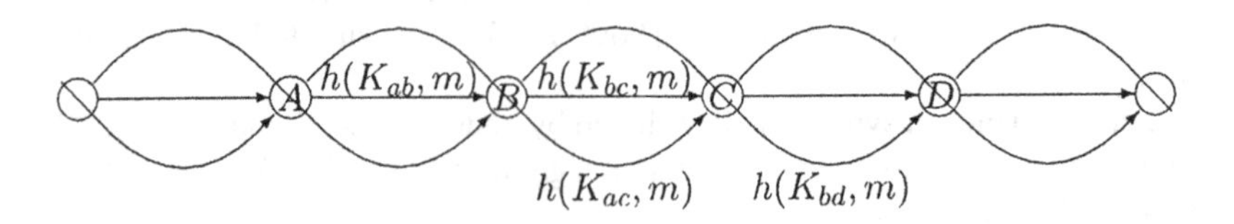

Figure 2.2. The Canvas protocol

Two adversarial nodes A and C separated by a honest node B can then collude to change a forwarded message m to $\tilde{m}$. The attack in Fig. 2.3 targets a node that can be reached in one hop from one of the adversarial nodes and in two hops from the other.

1 Adversary A forwards message m to B, naming C as the next node and including $h(K_{ae}, \tilde{m})$ in place of the authenticator $h(K_{ac}, m)$; E has to be a node that can be reached in one hop from C and in two hops from A.

2 Node B successfully checks the authenticators for m, names D as the next node, and forwards $h(K_{ae}, \tilde{m})$ unchecked.

3 Adversary C receives m from B, changes it according to the pre-arranged strategy to $\tilde{m}$, generates authenticators for the modified message, and forwards those together with $h(K_{ae}, \tilde{m})$ to E.

4 Node E receives the modified message $\tilde{m}$ with valid authenticators from A and C and accepts it as genuine.

This attack could be prevented if E knows about valid routes in the network. By assumption, A and C are not direct neighbours so messages could not arrive along the route $A \rightarrow C \rightarrow E$. However, this would constitute yet another change in assumptions. In our original set-up, nodes only store keys for some neighbours but have no further information about the network topology. We can only tell in concrete applications which assumptions about the environment are really justified.

4.5 Complexity – key insertion attacks

Most protocols studied in the academic literature on protocol analysis are not complex at all, nor are the attacks that have been found. Admittedly, even for simple protocols security proofs can become quite complex. True complexity, however, is met outside the traditional playing ground of Alice & Bob

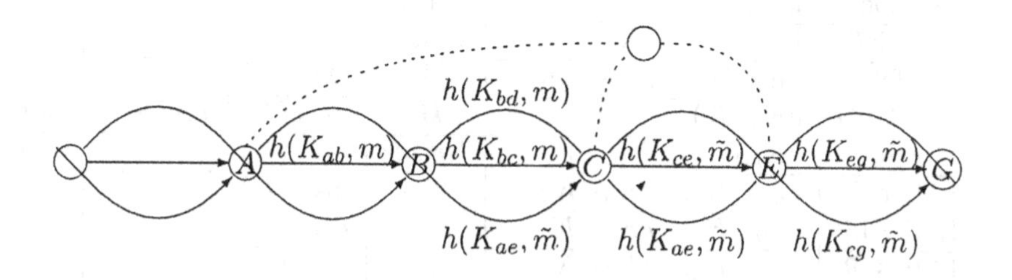

Figure 2.3. An "attack" on the Canvas protocol; dotted lines indicate unused links.

protocols. For illustration, we sketch a key insertion attack on a key management module described in [Longley and Rigby, 1992].

Key management modules are used in the financial sector for executing sensitive cryptographic operations. Only encrypted keys are allowed outside the module. Keys for different types of applications have tags defining their intended use. Our example has data encrypting keys (tags *kds* and *kdr* for sending and receiving encrypted data), key encrypting master keys (tags *kis*, *kir*), and change keys for re-encrypting keys (tag *kc*). Table 2.1 gives the functions provided by the module.

In a *key insertion attack* the adversary tries to get a key K of her choice encrypted under the master key K_m. The adversary can call all functions of the module and may have access to encrypted keys that have been legally exported from the module. It is a non-trivial exercise to construct such an attack, or to prove that a system is not vulnerable in this way. Key insertion attacks have received renewed attention in recent times [Bond and Anderson, 2001].

With the set of functions given in this case study, an attack can be mounted [Longley and Rigby, 1992]. The adversary starts with an encrypted but unknown key $eK_m(U)$ with tag *kc* and a random data block interpreted as $eU(X)$. In the attack the value X remains unknown to the adversary. The attack yields $eK_m(K)$ with tag *kis*:

1 $eK_m(X)[kc] \leftarrow \text{EMKKC}(eK_m(U), eU(X))$

2 $eK_m(X)[kis] \leftarrow \text{EMKKIS}(eK_m(U), eU(X))$

3 $eK_m(K')[kds], eX(K')[kdr] \leftarrow \text{KEYGEN}(eK_m(X))$

4 $eK_m(K')[kc] \leftarrow \text{EMKKC}(eK_m(X), eX(K'))$

5 $eK'(K)[kc] \leftarrow \text{SECENC}(K, eK_m(K'))$

6 $eK_m(K)[kis] \leftarrow \text{EMKKIS}(eK_m(K'), eK'(K))$

function	input	tags	output	tags
SECENC	DATA, $eK_m(K_D)$	-, *kds*	eK_D(DATA)	-
SECDEC	eK_D(DATA), $eK_m(K_D)$	-, *kdr*	DATA	-
KEYGEN	$eK_m(K)$	*kis*	$eK_m(K_G), eK(K_G)$	*kds, kdr*
RTMK	$eK_m(K_1), eK_1(K_2),$	*kir, kdr*	$eK_m(K_2)$	*kdr*
EMKKC	$eK_m(K_M), eK_M(K)$	*kc*, -	$eK_m(K)$	*kc*
EMKKIS	$eK_m(K_M), eK_M(K)$	*kc*, -	$eK_m(K)$	*kis*
EMKKIR	$eK_m(K_M), eK_M(K)$	*kc*, -	$eK_m(K)$	*kir*

Table 2.1. Functions of a key management module

The attack was found by coding the functions in Prolog and using unification to check whether the goal $eK_m(K)$ can be reached. There is no guarantee that this process will terminate. There is scope for research in dedicated search methods that work faster or give stronger guarantees.

Key insertion attacks can also be viewed as a particular instance of privilege escalation. Code based access control where code segments can *require* and *assert* privileges, as used in the .NET framework [La Macchia et al., 2002], raises similar issues. In these settings, there is also potential complexity in the formulation of security goals as we may have to account for a large number of privileges.

5. Conclusions and Challenges

There are common themes to the first four case studies. The Dolev-Yao model falls short of capturing all aspects of the situation we have to model and the choice of security goals needs careful consideration. Authentication still keeps coming up with surprises. More discussions about the various definitions of authentication and their relationships can be found e.g. in [Gollmann, 2003b, Gollmann, 1996, Gollmann, 2003a]. The standard reference for the current cryptologic terminology is [Menezes et al., 1997].

New attacks on established protocols are less the fruit of sophisticated analysis than the results of changes in assumptions about the protocol goals or the environment. Often, once the desired security properties and the assumptions about the environment are made explicit, the attacks are quite obvious and would have been found sooner or later by direct inspection. Of course, one advantage of tool support for protocol verification is that the attacks are found sooner rather than later.

Our understanding of security concepts tends to rely on features of the environment the concepts emerged from. When the environment changes familiar attitudes to security may in fact become misleading. Hence, efforts towards formal protocol analysis might be justified not so much by the complexity of the problems being tackled but by the fact that formal specifications encourage us to clarify and justify our assumptions. In contrast, analysis with off-the-shelf properties runs the danger of producing results that are irrelevant for the problem at hand.

Rules of thumb for the design of security protocols [Abadi and Needham, 1994] should equally be treated with caution. On one hand, they may be based on aspects that were taken for granted at the time the rules were proposed but in hindsight turn out to be specific features of the applications discussed at that time. On the other hand, they may be overly cautious and dissuade us from exploiting specific features of the application at hand.

Formal protocol analysis amounts to *symbolic debugging* more than to protocol verification. Even provably secure schemes can be, and have been broken outside the model they have been analysed in. When protocols are adapted for a new environment, it may well occur that some of the assumptions underpinning the original security analysis are no longer valid, and that a familiar protocol is then no longer secure.

Challenges in protocol design and analysis come from novel application areas like mobility or ubiquitous computing. There, standard but inappropriate assumptions about the nature of security have to be identified and eliminated. Formal analysis needs agile methodologies. Methods and tools that analyze standard properties under standard assumptions are of little help. They have, however, a role to play in supporting security-unaware application writers who are dealing with well-understood security issues.

Challenges related to the complexity of the problem arise in multi-party protocols such as group key exchange, in multi-stage protocols such as optimistic fair exchange protocols, or in settings where multiple functions can interact as in our last case study.

Notes

1. Source [Vogt, 2003] refers to message authentication, but in [Menezes et al., 1997] this term appears as a synonym for data origin authentication.

References

[Abadi and Needham, 1994] Abadi, Martín and Needham, Roger (1994). Prudent engineering practice for cryptographic protocols. In *Proceedings of the 1994 IEEE Symposium on Research in Security and Privacy*, pages 122–136.

[Aura et al., 2002] Aura, Tuomas, Roe, Michael, and Arkko, Jari (2002). Security of Internet location management. In *Proceedings of the 18th Annual Computer Security Applications Conference*, pages 78–87.

[Bird et al., 1992] Bird, Ray, Gopal, Inder, Herzberg, Amir, Janson, Phil, Kutten, Shay, Molva, Refik, and Yung, Moti (1992). Systematic design of two-party authentication protocols. In Feigenbaum, J., editor, *Advances in Cryptology – CRYPTO'91, LNCS 576*, pages 44–61. Springer Verlag.

[Bond and Anderson, 2001] Bond, Mike and Anderson, Ross (2001). API-level attacks on embedded systems. *IEEE Computer*, 34(10):67–75.

[Burrows et al., 1990] Burrows, Michael, Abadi, Martín, and Needham, Roger (1990). A logic of authentication. *DEC Systems Research Center*, Report 39.

[Dolev and Yao, 1983] Dolev, Danny and Yao, Andrew C. (1983). On the security of public key protocols. *IEEE Transactions on Information Theory*, IT-29(2):198–208.

[Ellison et al., 1999] Ellison, Carl M., Frantz, Bill, Lampson, Butler, Rivest, Ron, Thomas, Brian M., and Ylonen, Tatu (1999). *SPKI Certificate Theory*. RFC 2693.

[Gollmann, 1996] Gollmann, Dieter (1996). What do we mean by entity authentication? In *Proceedings of the 1996 IEEE Symposium on Security and Privacy*, pages 46–54.

[Gollmann, 2003a] Gollmann, Dieter (2003a). Analysing security protocols. In Abdallah, A. E., Ryan, P., and Schneider, S., editors, *Formal Aspects of Security, LNCS 2629*, pages 71–80. Springer Verlag.

[Gollmann, 2003b] Gollmann, Dieter (2003b). Authentication by correspondence. *IEEE Journal on Selected Areas in Communications*, 21(1):88–95.

[Gong, 1999] Gong, Li (1999). *Inside Java 2 Platform Security*. Addison-Wesley, Reading, MA.

[La Macchia et al., 2002] La Macchia, Brian A., Lange, Sebastian, Lyons, Matthew, Martin, Rudi, and Price, Kevin T. (2002). *.NET Framework Security*. Addison-Wesley Professional, Boston, MA.

[Lampson et al., 1992] Lampson, Butler, Abadi, Martín, Burrows, Michael, and Wobber, Edward (1992). Authentication in distributed systems: Theory and practice. *ACM Transactions on Computer Systems*, 10(4):265–310.

[Longley and Rigby, 1992] Longley, D. and Rigby, S. (1992). An automatic search for security flaws in key management schemes. *Computers & Security*, 11(1):75–89.

[Lowe, 1995] Lowe, Gavin (1995). An attack on the Needham-Schroeder public-key authentication protocol. *Information Processing Letters*, 56(3):131–133.

[Lowe, 1996] Lowe, Gavin (1996). Breaking and fixing the Needham-Schroeder public-key protocol using FDR. In T. Margaria, B. Steffen, editor, *Proceedings of TACAS, LNCS 1055*, pages 147–166. Springer Verlag.

[Mäki and Aura, 2002] Mäki, Silja and Aura, Tuomas (2002). Towards a survivable security architecture for ad-hoc networks. In et al., B. Christiansen, editor, *Security Protocols, 9th International Workshop, Cambridge, LNCS 2467*, pages 63–73. Springer Verlag.

[Menezes et al., 1997] Menezes, Alfred J., van Oorschot, Paul C., and Vanstone, Scott A. (1997). *Handbook of Applied Cryptography*. CRC Press, Boca Raton, FA.

[Mitchell and Pagliusi, 2003] Mitchell, Christopher J. and Pagliusi, Paolo S. (2003). Is entity authentication necessary? In et al., B. Christiansen, editor, *Security Protocols, 10th International Workshop, Cambridge, LNCS 2845*, pages 20–33. Springer Verlag.

[Needham, 2000] Needham, Roger (2000). Keynote address: The changing environment (transcript of discussion). In et al., B. Christiansen, editor, *Security Protocols, 7th International Workshop, Cambridge, LNCS 1796*, pages 1–5. Springer Verlag.

[Needham and Schroeder, 1978] Needham, Roger M. and Schroeder, Michael D. (1978). Using encryption for authentication in large networks of computers. *Communications of the ACM*, 21:993–999.

[Ryan et al., 2001] Ryan, Peter, Schneider, Steve, Goldsmith, Michael, Lowe, Gavin, and Roscoe, Bill (2001). *Modelling and Analysis of Security Protocols*. Addison-Wesley, Harlow, England.

[Schneider, 1996] Schneider, Steve (1996). Security properties and CSP. In *Proceedings of the 1996 IEEE Symposium on Security and Privacy*, pages 174–187.

[Schneider, 1998] Schneider, Steve (1998). Verifying authentication protocols in CSP. *IEEE Transactions on Software Engineering*, 24(9):741–758.

[Vogt, 2003] Vogt, Harald (2003). Integrity preservation for communication in sensor networks. private communications.

Chapter 3

PRIVATE MATCHING

Yaping Li
UC Berkeley

J. D. Tygar
UC Berkeley

Joseph M. Hellerstein
UC Berkeley
Intel Research Berkeley

Abstract Consider two organizations that wish to privately match data. They want to find common data elements (or perform a join) over two databases without revealing private information. This was the premise of a recent paper by Agrawal, Evfimievski, and Srikant. We show that Agrawal et al. only examined one point in a much larger problem set and we critique their results. We set the problem in a broader context by considering three independent design criteria and two independent threat model factors, for a total of five orthogonal dimensions of analysis.

Novel contributions include a taxonomy of design criteria for private matching , a secure data ownership certificate that can attest to the proper ownership of data in a database, a set of new private matching protocols for a variety of different scenarios together with a full security analysis. We conclude with a list of open problems in the area.

1. Introduction

Agrawal, Evfimievski, and Srikant recently presented a paper [Agrawal et al., 2003] that explores the following *private matching* problem: two parties each have a database and they wish to determine common entries without revealing any information about entries only found in one database. This paper has generated significant interest in the research community and technical press. While the Agrawal/Evfimievski/Srikant (AgES) protocol is correct within in its assumptions, it is not robust in a variety of different scenarios. In fact, in many likely scenarios, the AgES protocol can easily be exploited to obtain a great deal of information about another database. As we discuss in this paper, the private matching problem has very different solutions depending on assumptions about the different parties, the way they interact, and cryptographic mechanisms available. Our paper discusses flaws in the AgES protocol, presents that protocol in the context of a framework for viewing private matching and a family of possible protocols, and gives a number of new techniques for addressing private matching, including a flexible powerful Data Ownership Certificate that can be used with a variety of matching protocols.

The private matching problem is a practical, constrained case of the more general (and generally intractable) challenge of *secure multi-party computation* . Private set matching is a simple problem that is at the heart of numerous data processing tasks in a variety of applications. It is useful for relational equijoins and intersections, as well as for full-text document search, cooperative web caching, preference matching in online communities, and so on. Private matching schemes attempt to enable parties to participate in such tasks without worrying that information is leaked.

In this paper we attempt a holistic treatment of the problem of two-party private matching. We lay out the problem space by providing a variety of possible design goals and attack models. We place prior work in context, and present protocols for points in the space that had been previously ignored. We also point out a number of additional challenges for future investigation.

1.1 Scenarios

We begin our discussion with three scenarios, which help illustrate various goals of a private matching protocol.

Our first scenario comes from multi-party customer relationship management in the business world. Two companies would like to identify their common customers for a joint marketing exercise, without divulging any additional customers. In this scenario, we would like to ensure that (a) neither party learns more than their own data and the answer (and anything implied by the pair), and (b) if one party learns the results of the match, both parties should learn it. Agrawal, et al. discuss a special instance of this case in their work [Agrawal et al., 2003], which they call *semi-honesty* , after terminology used in secure multi-party literature [Goldreich, 2002]. In particular, the two companies are assumed to honestly report their customer lists (or, more generally, the lists they wish to intersect), but may try otherwise to discover additional information about the other's customer list. The semi-honest scenario here rests on the presumption that a major corporation's publicity risk in being detected lying outweighs its potential benefit in one-time acquisition of competitive information. Below, we comment further on difficulties raised by this notion of semi-honesty.

In many cases, we do not desire symmetric exchange of information. As a second example, consider the case of a government agency that needs to consult a private database. Privacy and secrecy concerns on the part of the government agency may lead it to desire access to the private database without revealing any information about the nature of the query. On the other hand, the database owner may only want to release information on a "need-to-know" basis: it may be required by law to release the answers to the specific query, but may be unwilling to release any other information to the government. In short, a solution to the situation should enable the government to learn only the answer to its query, while the database owner will learn nothing new about the government. In this asymmetric scenario, we need a different choice than (b) above.

Finally, we consider a scenario that could involve anonymous and actively dishonest parties. Online auction sites are now often used as a sales channel for small and medium-sized private businesses. Two competing sellers in an online auction site may wish to identify and subsequently discuss the customers they have in common. In this case, anonymity of the sellers removes the basis for any semi-honesty assumption, so guaranteed mechanisms are required to prevent one party from tricking the other into leaking information.

Each of these examples has subtly different design requirements for a private matching protocol. This paper treats these examples by systematically exploring all possible combinations of security requirements along a number of independent design criteria.

1.2 Critique of AgES

In their paper [Agrawal et al., 2003], Agrawal, Evfimievski, and Srikant consider the first scenario listed above, building on an earlier paper by Huberman et al. [Huberman et al., 1999]. Here is an informal summary of the AgES Set Intersection Protocol result; we discuss it more formally below in Section 3.

Agrawal, et al. suggest solving the matching problem by introducing a pair of encryption functions E (known only to A) and E' (known only to B) such that for all x, $E(E'(x)) = E'(E(x))$. Alice has customer list A and Bob has customer list B. Alice sends Bob the message $E(A)$; Bob computes and then sends to Alice the two messages $E'(E(A))$ and $E'(B)$. Alice then applies E to $E'(B)$, yielding (using the commutativity of E and E') these two lists: $E'(E(A))$ and $E'(E(B))$. Alice computes $E'(E(A)) \cap E'(E(B))$. Since

Alice knows the order of items in A, she also knows the the order of items in $E'(E(A))$ and can quickly determine $A \cap B$.

Two main limitations are evident in this protocol. First, it is *asymmetric*: if we want both parties to learn the answer, we must trust Alice to send $A \cap B$ to Bob. This asymmetry may be acceptable or even desirable in some scenarios, but may be undesirable in others.

Second, we find the AgES assumption of *semi-honesty* to be hard to imagine in a real attack scenario. Any attacker who would aggressively decode protocol messages would presumably not hesitate to "spoof" the contents of their queries. If we admit the possibility of the attacker spoofing queries, then the AgES protocol is not required; a simpler hash-based scheme suffices. In this scheme (also suggested by Agrawal, et al.) the two parties hash the elements of their lists $h(A)$ and $h(B)$ and then compute the intersection of those two lists of hashes. Later in this paper, we augment this hash-based protocol with an additional mechanism to prevent spoofing as well.

1.3 A broader framework

Below, we consider a broader framework for thinking about private matching.

First, we break down the protocol design space into three independent criteria :

Design criteria

- protocols that leak no information (*strong*) vs. protocols that leak some information (*weak*)
- protocols that protect against spoofed elements (*unspoofable*) vs. protocols that are vulnerable (*spoofable*).
- symmetric release of information vs. asymmetric release (to only one party).

We will also consider two different dimensions for threat models:

Threat models

- semi-honest vs. malicious parties
- small vs. large data domains

We discuss the design criteria in more detail in the next section and cover the threat models below in Section 3.

2. Problem Statement

We define the *private matching* problem between two parties as follows. Let the two parties Alice and Bob have respective sets A and B of objects in some domain D. Suppose Alice wants to pose a matching query $Q \subseteq D$ to Bob. We call Alice the *initiator* of the query and Bob the *recipient* of the query. We say Q is *valid* if $Q \subseteq A$ and *spoofed* otherwise. A *matching* computes $P = Q \cap B$ or $\perp$; note that $\perp$ is a message distinguishable from the set $\emptyset$, and can be thought of as a warning or error message.

We elaborate upon the three design criteria for private matching described in the previous section:

- We say that a matching protocol is *strong* if any party can learn only: P, any information that can be derived from P and this party's input to the protocol, the size of the other party's input, and nothing else; otherwise the protocol is *weak* with respect to the additional information learnable.

- We define a matching protocol to be *unspoofable* if it returns $\perp$ or $Q \cap A \cap B$ for all spoofed Q. Otherwise it is *spoofable*.

- We say that a matching protocol is *symmetric* if both parties will know the same information at any point in the protocol. Otherwise it is *asymmetric*.

For each of these three dimensions, a bit more discussion is merited. We begin with the *strong/weak* dichotomy. After executing a protocol, a party can derive information by computing functions over its input to the protocol and the protocol's output. An example of such derived information is that a party can learn something about what is *not* in the other party's set, by examining its input and the query result. Since any information that can be computed in this way is an unavoidable consequence of matching, we use P to denote both P and the derived information throughout our paper. Note that weak protocols correspond to the notion of semi-honesty listed above — weak protocols allow additional information to be leaked, and only make sense when we put additional restrictions on the parties — typically, that they be semi-honest. In contrast, strong protocols allow malicious parties to exchange messages. Note that we allow the size of a party's input to be leaked; the program of each party in a protocol for computing a desired function must either depend only on the length of the party's input or obtain information on the counterpart's input length [Goldreich, 2002].

For the *spoofable/unspoofable* dimension, there are scenarios where a protocol that is technically spoofable can be considered effectively to be unspoofable. To guarantee that a protocol is unspoofable, it requires the protocol to detect spoofed queries. Given such a mechanism, either of the following two responses are possible, and maintain the unspoofable property: (a) returning $\perp$, or (b) returning $Q \cap A \cap B$. When a party lacks such a detection mechanism, it cannot make informed decision as when to return $\perp$. However, in some situations, the party may be expected to return the set $Q \cap A \cap B$ with high probability, regardless of whether the query is spoofed or not. This may happen when it is very difficult to spoof elements. We will give an example of this scenario later.

It is also useful to consider the the issue of symmetry vs. asymmetry for the threat models covered in Section 3. In the semi-honest model, parties follow the protocols properly, and so symmetry is enforced by agreement. However, in a malicious model, the parties can display arbitrary adversarial behavior. It is thus difficult to force symmetry, because one party will always receive the results first. (A wide class of cryptographic work has revolved around "fair exchanges" in which data is released in a way that guarantees that both parties receive it, but it is not clear if those concepts could be efficiently applied in the private matching application.)

2.1 Secure multi-party computation

The private matching problem is a special case of the more general problem from the literature called *secure multi-party computation*. We now give a brief introduction to secure multi-party computation in the hope of shedding light on some issues in private matching. In a secure m-party computation, the parties wish to compute a function f on their m inputs. In an *ideal model* where a trusted party exists, the m parties give their inputs to the trusted party who computes f on their inputs and returns the result to each of the parties. The results returned to each party may be different. This ideal model captures the highest level of security we can expect from multi-party function evaluation [Canetti, 1996]. A secure multi-party computation protocol emulates what happens in an ideal model. It is well-known that no secure multi-party protocol can prevent a party from cheating by changing its input before a protocol starts [Goldreich, 2002]. Note however, that this cannot be avoided in an ideal model either. Assuming the existence of trapdoor permutations, one may provide secure protocols for any two-party computation [Yao., 1986] and for any multi-party computation with honest-majority [Goldreich et al., 1987]. However, multi-party computations are usually extraordinarily expensive in practice, and impractical for real use. Here, our focus is on *highly efficient* protocols for private matching, which is both tractable and broadly applicable in a variety of contexts.

3. Threat Models

We identify two dimensions in the threat model for private matching. The first dimension concerns the domain of the sets being matched against. A domain can be *small*, and hence vulnerable to an exhaustive search attack , or *large*, and hence not vulnerable to an exhaustive search attack.

If a domain is small, then an adversary Max can enumerate all the elements in that domain and make a query with the entire domain to Bob. Provided Bob answers the query honestly, Max can learn the entirety of Bob's set with a single query. A trivial example of such a domain is the list of Fortune 500 companies; but note that there are also somewhat larger but tractably small domains like the set of possible social security numbers.

A large uniformly distributed domain is not vulnerable to an exhaustive search attack. We will refer to this type of domain simply as *large* in this paper. An example of such a domain is the set of all RSA keys of a certain length. If a domain is large, then an adversary is limited in two ways. First, the adversary cannot enumerate the entire domain in a reasonable single query, nor can the adversary repeatedly ask smaller queries to enumerate the domain. In this way the adversary is prevented from mounting the attack described above. Second, it is difficult for her to query for an arbitrary individual value that another party may hold, because each party's data set is likely to be a negligible-sized subset of the full domain.

The second dimension in the threat model for private matching captures the level of adversarial misbehavior. We distinguish between a semi-honest party and a malicious party [Goldreich, 2002]. A semi-honest party is honest on its query or data set and follows the protocol properly with the exception that it keeps a record of all the intermediate computations and received messages and manipulates the recorded messages in an aggressively adversarial manner to learn additional information.[1] A malicious party can misbehave in arbitrary ways: in particular, it can terminate a protocol at arbitrary point of execution or change its input before entering a protocol. No two-party computation protocol can prevent a party from aborting after it receives the desired result and before the other party learns the result. Also no two-party computation protocol can prevent a party from changing its input before a protocol starts.

Hence we have four possible threat models: a semi-honest model with a small or large domain, and a malicious model with a small or large domain. In the rest of the paper, we base our discussion of private matching protocols in terms of these four threat models.

3.1 Attacks

In this section we enumerate a number of different attacks that parties might try to perform to extract additional information from a database. In the scenarios below, we use the notation A and B to denote parties, and A is trying to extract information from B's database.

- **Guessing attack:** In this attack, the parties do not deviate from the protocol. However, A attempts to guess values in B's database and looks for evidence that those values occur in B's database. Typically, A would guess a potential value in B's database, and then look for an occurrence of the hash in B's database. Alternatively, A could attempt to decrypt values in a search for an encrypted version of a particular potential value in B's database (following the pattern in the AgES protocol.) Because of the limitations of this type of attack, it is best suited when the domain of potential values is small. (A variant of this attack is to try all potential values in the domain, an *exhaustive search attack.*)

- **Guess-then-spoof attack:** In this attack, the parties deviate from the protocol. As in the guessing attack, A generates a list of potential values in B's database. In the spoofing attack , A runs through the protocol pretending that these potential values are already in A's database. Thus A will compute hashes or encrypt, and transmit values as if they really were present in A's database. Because this attack involves a guessing element, it is also well suited for small domains of potential database values (e.g. social security numbers, which are only 10 digits long).

- **Collude-then-spoof attack:** In this attack, A receives information about potential values in B's database by colluding with outside sources. For example, perhaps A and another database owner C collude by exchanging their customer lists. A then executes a spoofing attack by pretending that these entries are are already on its list. As in guess-then-spoof attack, A computes hashes or encrypts, and transmits values as if they were really present in A's database. Since A is deriving its information from third party sources in this attack, it is suited for both small and large domains of potential database values. (N.B.: we group both the guess-then-spoof attack and the collude-then-spoof attack together as instances of *spoofing attacks*. Spoofing attacks occur in the malicious model; in the semi-honest model they can not occur.)

- **Hiding attacks:** In a hiding attack, A only presents a subset of its customer list when executing a matching protocol, effectively hiding the unrevealed members. This paper does not attempt to discuss defenses against hiding attacks.

Although we would like to prevent all collusion attacks involving malicious data owners, there are limits to what we can accomplish. For example, if Alice and Bob agree to run a matching protocol, nothing can prevent Bob from simply revealing the results to a third party Charlie. In this case, Bob is acting as a proxy on behalf of Charlie, and the revelation of the results occurs out-of-band from the protocol execution. However, we would like to resist attacks where Bob and Charlie collude to disrupt the protocol execution or use inputs not otherwise available to them.

4. Terminology and Assumptions

We begin by assuming the existence of *one-way collision resistant hash functions* [Menezes et al., 1996] . A hash function $h(\cdot)$ is said to be one-way and collision resistant if it is difficult to recover M given $h(M)$, and it is difficult to find $M' \neq M$ such that $h(M') = h(M)$. Let $\text{SIGN}(\cdot, \cdot)$ be a public key signing function which takes a secret key and data and returns the signature of the hash of the the data signed by the secret key. Let $\text{VERIFY}(\cdot, \cdot, \cdot)$ be the corresponding public key verification function which takes a public key, data, and a signature and returns **true** if the signature is valid for the data and **false** otherwise. For shorthand, we denote $\{P\}_{sk}$ as the digital signature signed by the secret key sk on a plaintext P. The function $isIn(\cdot, \cdot)$ takes an element and a set and returns **true** if the element is in the set and **false** otherwise.

The *power function* f : Key$\mathcal{F}$ $\times$ Dom$\mathcal{F}$ $\rightarrow$ Dom$\mathcal{F}$ where f defined as follows:

$$f_e(x) \equiv x^e \bmod p$$

is a *commutative encryption* [Agrawal et al., 2003]:

- The powers commute:

$$(x^d \bmod p)^e \bmod p \equiv x^{de} \bmod p \equiv (x^e \bmod p)^d \bmod p$$

- Each of the powers f_e is a bijection with its inverse being $f_e^{-1} \equiv f_{e^{-1} \bmod q}$.

where both p and $q = (p-1)/2$ are primes.

We use the notation $e \stackrel{r}{\leftarrow} S$ to denote that element e is chosen randomly (using a uniform distribution) from the set S.

We assume there exists an encrypted and authenticated communication channel between any two parties.

1 Alice's local computation:

(a) $Q_h := \{h(q) : q \in Q\}$.

(b) $e_A \stackrel{r}{\leftarrow} Key\mathcal{F}$.

(c) $Q_{e_A} := \{f_{e_A}(q_h) : q_h \in Q_h\}$.

2 Bob's local computation:

(a) $B_h := \{h(b) : b \in B\}$,

(b) $e_B \leftarrow_r Key\mathcal{F}$.

(c) $B_{e_B} := \{f_{e_B}(b_h) : b_h \in B_h\}$.

3 Alice→ Bob: Q_{e_A}.

4 Bob's local computation:
$Q_{e_A,e_B} := \{(q_{e_A}, f_{e_B}(q_{e_A})) : q_{e_A} \in Q_{e_A}\}$.

5 Bob→ Alice: B_{e_B}, Q_{e_A,e_B}.

6 Alice's local computation:

(a) $Q'_{e_A,e_B} := \emptyset, P := \emptyset$

(b) $B_{e_B,e_A} := \{f_{e_A}(b_{e_B}) : b_{e_B} \in B_e\}$.

(c) For every $q \in Q$, we compute $q_{e_A} = f_{e_A}(h(q))$, and find the pair $(q_{e_A}, q_{e_A,e_B}) \in Q_{e_A,e_B}$; given this we let $Q'_{e_A,e_B} := Q'_{e_A,e_B} \cup \{(q, q_{e_A,e_B})\}$.

(d) For every $(q, q_{e_A,e_B}) \in Q'_{e_A,e_B}$, $\mathbf{if\ isIn(q_{e_A,e_B}, B_{e_B,e_A}),\ then P := P \cup \{q\}}$.

7 Output P.

Figure 3.1. AgES protocol

5. Techniques

We present three matching protocols in this section: the trusted third party protocol , the hash protocol , and the AgES protocol [Agrawal et al., 2003]. In the next section, we describe a data ownership certificate mechanism that can be combined with all three protocols to *despoof* all of the original protocols even in threat models with small domains.

5.1 Trusted Third Party Protocol (TTPP)

Suppose Alice and Bob trust a third party Trudy. Alice and Bob can compute their private matching through Trudy. Alice (resp. Bob) sends her query Q (resp. his data set B) to Trudy, and Trudy computes the intersection P of the two sets. Trudy then returns the result to both parties in the symmetric case, or to one of the parties in the asymmetric case.

We discuss the security of the TTPP in Section 7.

5.2 Hash Protocol (HP)

In this section, we present a *Hash Protocol* that do not require a trusted third party . In the hash protocol, Alice sends Bob her set of hashed values. Bob hashes his set with the same hash function, and computes the intersection of the two sets. Bob may send Alice the result based on their prior agreement.

We discuss the security of the hash protocol in Section 7.

5.3 The AgES protocol

We gave a summary of the AgES protocol in Section 1.2. Now we present the complete version of the protocol in Figure 3.1. For consistency we adapt this protocol to our notation, but the essence of the protocol remains the same as the original paper.

We discuss the security of the AgES protocol in Section 7.

6. Data Ownership Certificate (DOC)

An especially difficult attack for private matching to handle is the spoofing problem. In this section, we propose a new approach to address spoofing: the use of *Data Ownership Certificates.* The idea is to have the creator of data digitally sign the data in a particular way so that parties that control databases that include the data can not spoof data. For example, consider the case of two companies each of which wants to find out as much as possible about the other's customer list. If one of the companies has access to a list of all residents in a particular area, a straightforward spoofing attack is quite simple — it could simply create false entries corresponding to a set of the residents. If any of those residents were on the other company's customer list, private matching would reveal their membership on that list. However, if the companies are obligated to provide digitally signed entries, this type of spoofing would be eliminated: neither of the companies would be able to falsify entries.

The above sketch is not sufficient, however, because it still leaves open the possibility that corrupt companies could broker in digitally signed data entries. For example, if customer E is a legitimate customer of firm F, we would have the possibility that F might try to trade or sell G's digitally signed entry to A. Then A would be able to falsely claim that G was a customer and during private matching, steal information through a spoofing attack. Below, we discuss an architecture for data ownership certificates that resists both regular spoofing attacks and colluding spoofing attacks .

Data Ownership Certificates do require more work on the part of individuals creating data, and they are probably only practical in the case of an individual who uses his or her computer to submit information to a database. Despite the extra work involved, we believe that data ownership certificates are not far-fetched. In particular, the European Union's Privacy Directive [Parliament, 1995] requires that individuals be able to verify the correctness of information about them and control the distribution of that information. Data Ownership Certificates give a powerful technical mechanism supporting that distribution. Similarly, Agrawal, Kieran, Srikant, and Xu have recently argued for a type of "Hippocratic Database" that would provide similar functionality [Agrawal et al., 2002]. Data Ownership Certificates would work well with these Hippocratic Databases.

Now we begin a formal presentation of Data Ownership Certificates (DOC). A Data Ownership Certificate is an authorization token which enables a set owner to prove it is a legitimate owner of some particular data. The first goal of the DOC is to prevent spoofing in a small domain. Data Ownership Certificates prevent spoofing by 'boosting" the size of the small domain D to a larger domain $D \times S$, where S is the domain of the DOCs. The intuition is that by expanding the domain, DOCs make the probability of guessing a correct value negligible in the cryptographic sense and protect database owners from guess-then-spoof attacks. Now, if an attacker wants to spoof a particular value, e.g. John's information, the attacker needs to correctly guess the associated DOC as well.

A second goal of Data Ownership Certificates is access control. A DOC is essentially a non-transferable capability issued by the originator of data to a database owner. We refer to the originators of data as *active entities*. We say that an active entity E *authorizes* a set owner O *sharing access* to its information d when E issues O a DOC C_d^O for d. Ideally, a common element between two databases should be discovered only when both databases have been authorized with DOCs by the corresponding active entity for that element. More precisely, we require two security properties from Data Ownership Certificates:

- Confidentiality: If Bob is not an authorized owner of d, Bob should not be able to learn that Alice possesses d if he runs a matching protocol directly with Alice.
- Authenticity: If Bob is not an authorized owner of d and Alice is an authorized owner of d, Bob should not be able to pollute Alice's matching result, i.e., Bob cannot introduce d into the matching result.

We find that confidentiality is difficult to achieve. We thought of two approaches to do the access control. First, Alice checks whether Bob has the authorization before she gives an element v to Bob. It seems essential that Alice obtains some knowledge k that links the access controlled object v to requester Bob before granting the access. This requester-specific knowledge k reveals at least partial information of what element Bob has. It is then only fair that Bob checks for Alice's permission to access k. This leads to an infinite reduction. Second, Alice can give Bob a box which contains John's information d. The box is locked by John. Bob can only open the box if he has the key. This implies that John uses a lock for which he knows Bob has the key. This kind of precomputation on John's part is not desirable. We leave this as an open problem for future work and we relax our requirement for access control in this paper. We propose a third solution that allows two parties to learn their common element d if both of them have d and some *common nonce* for d instead of some requester specific access token. We refer to the goal of DOC as *reduced confidentiality requirement.*

6.1 Our instantiation of DOCs

Our instantiation of Data Ownership Certificates consists of two parts: a common nonce (random string) and an ownership attestation component. The common nonce serves the purpose of both boosting the domain and satisfying the reduced confidentiality requirement. The ownership attestation component satisfies the authenticity requirement.

A Data Ownership Certificate C has the form of $\langle pk, n, \sigma \rangle$. Each active entity E maintains three keys k_1, sk, and pk. For each piece of information d originating from E, E generates a unique $n = G(k_1||d)$ where $G(\cdot)$ is a pseudo-random number generator and $||$ is the concatenation function. Assume that the output n of $G(\cdot)$ is l bits long and $G(\cdot)$ is cryptographically secure, then by the birthday paradox, one needs to guess approximately $\sqrt{2^l}$ numbers to have one of them collide with n. If l is large enough, say 1024, then guessing the correct n is hard. This nonce n will be used in matching protocols instead of the original data d.

When E submits d to some database A, it generates a signature $\sigma = \{d||A\}_{sk}$ where A is the unique ID of the database. The signature does not contain the plaintext information d or A, however anyone knowing the public pk and the plaintext information d and A may verify that A is indeed an authorized owner of d by verifying the authenticity of σ using pk.

6.2 Certified matching protocols

In this section, we describe the integration of Data Ownership Certificates with the proposed protocols from Section 5.

We assume that each set element in database A is a pair $\langle d, C\rangle$ of data and a Data Ownership Certificate $C = \langle pk, n, \sigma\rangle$ where $\sigma = \{d||A\}_{sk}$. The owner of database A now runs a matching protocol with n instead of d as the data.

6.2.1 Certified Trusted Third Party Protocol (CTTPP). We describe how to use Data Ownership Certificates to extend the Trusted Third Party Protocol . Let A (resp. B) be the ID of Alice (resp. Bob). The set that Alice (resp. Bob's) sends to Trudy contains elements in the form of (n_a, σ_a, pk_{n_a}) (resp. (n_b, σ_b, pk_{n_b})), i.e., triples of a common nonce, ownership attestation component, and the corresponding public key. The nonce n_a (resp. n_b) is associated with elements a (resp. b).

When Trudy finds a matching between two common nonces n_a and n_b, she compares the corresponding public keys pk_{n_a} and pk_{n_b}. If they are not the same, then it means that Alice and/or Bob spoofed the element and forged the corresponding certificate. Trudy cannot tell which is the case and she simply returns $\perp$ to both of them. If the corresponding public keys are the same, Trudy runs the verification algorithm on Alice's and Bob's ownership attestation component $\text{VERIFY}(pk_{n_a}, a||A, \sigma_a) = v_2$ and $\text{VERIFY}(pk_{n_b}, b||B, \sigma_b) = v_2$ to check whether Alice and/or Bob are authorized owners of the matching value. Trudy will find one of the following three cases to be true:

1 $v_1 = \mathbf{true}$ and $v_2 = \mathbf{true}$

2 $v_1 = \mathbf{true}$ and $v_2 \neq \mathbf{true}$ or vice versa

3 $v_1 \neq \mathbf{true}$ and $v_2 \neq \mathbf{true}$

If Trudy encounters case (1), then she concludes Alice and Bob are the authorized owners of the matching element. She adds the element to the result set and continues with the matching computation. We show why this is the case. Suppose only Bob is the authorized owner of the element associated with n_b. It is unlikely that Alice spoofs the common nonce n_a where $n_a = n_b$ as discussed in Section 6.1. Suppose Alice obtains n_a and the associated DOC for some other database owner, it is highly unlikely that Alice can generate a public/private key pair that is the same as the key pair for n_b. By symmetry, it is highly unlikely to be the case that Alice is the authorized owner of the element associated with n_a and Bob spoofs n_b or the public/private key pair. If (2) or (3) is the case, it implies Alice and/or Bob spoofed the nonce and an associated DOC or obtained her/his element from some other authorized owner(s) and spoofed a DOC. Trudy returns $\perp$ for this case.

If Alice (resp. Bob) did not pose a spoofed query and receives $\perp$ from Trudy, then she (resp. he) knows that the other party was not honest.

6.2.2 Certified Hash Protocol (CHP). The integration of data ownership certificates with the Hash Protocol is slightly different from that with the Trusted Third Party Protocol . We assume that Alice poses a query Q_h each element of which is in the form of $\langle h(n_a), \sigma_a \rangle$ where $\sigma_a = \{a||A\}_{sk_a}$.

Bob hashes each of his common nonces and checks if it matches one of $h(n_a)$. If he discovers a match between $h(n_a)$ and $h(n_b)$, then he assumes that the two corresponding ownership attestation components were signed by the same private key and does the following check. Bob first looks up his copy of the public key pk_b for n_b and checks if $\text{VERIFY}(pk_b, b||A, \sigma_a)$ returns $\mathtt{true}$. If it does return $\mathtt{true}$, it means that Alice is an authorized owner of b. Bob may add b to the result set P and continue with his matching computation. Otherwise Bob can conclude that Alice is not the authorized owner of b — she either obtained $h(n_a)$ and the corresponding certificate from some other authorized owner of a or she was able to guess $h(n_a)$ and forged the ownership attestation component. Bob cannot tell which was the case. Now Bob has the following two options: (a) returning $\perp$ to Alice, or (b) continuing with the matching computation but omitting b from the final result. Either way the modified protocol satisfies the security goal of being unspoofable and it enables parties to detect cheating.

We need to be careful about the usage of hash functions in the Certified Hash Protocol. Consider the following two scenarios. In the first scenario, assume that Alice, Bob, and Charlie are authorized owners of some customer John's information d. Imagine Alice executes the Certified Hash Protocol with Bob and Charlie and she receives data from Bob and Charlie. If Bob and Charlie use the same hash function, e.g. MD5 or SHA1 , then Alice may infer that all three of them have d after the protocol executions with Bob and Charlie respectively. Alice hashes her own copy of the nonce n_d associated with d and discovers n_b is in the sets that Bob and Charlies sends to her. The second scenario is that both Bob and Charlie are authorized owners of d but Alice is not. Furthermore, assume Alice does not have a copy of d and its DOC from some other authorized owner. In this case, Alice may infer that Bob and Charlie share some common information although she does not know what it is.

We propose using an HMAC in the Hash Protocol to prevent the inference problem in the second scenario. An HMAC is a keyed hash function that is proven to be secure as long as the underlying hash function has some reasonable cryptographic strength [Bellare et al., 1996]. An $\text{HMAC}_k(T) =$

$$h(k \oplus \text{ opad}, h(k \oplus \text{ipad}, T))$$

is a function which takes as inputs a secret key k and a text T of any length; "opad" and "ipad" are some predetermined padding. The output is an l-bit string where l is the output of the underlying hash function $h(\cdot)$.

Using HMAC in the Certified Hash Protocol avoids the problem in the second scenario as long as each pair of parties uses a different key every time they run the Certified Hash Protocol. This prevents adversaries from correlating elements from different executions of the Hash Protocol.

6.2.3 Certified AgES protocol (CAgES). We need to modify the AgES protocol in Figure 3.1 in three ways. First, both Alice and Bob hash and encrypt the common nonce instead of the actual data. Second, Bob returns pairs $\langle \sigma_b, f_{e_B}(h(n_b)) \rangle$ for each of his encrypted elements $f_{e_B}(h(n_b))$. Third, whenever there is a match, Alice verifies whether Bob is an authorized owner by checking the corresponding σ_b.

6.3 Homomorphic DOC (HDOC)

The data ownership certificate as proposed is limited in a way that it introduces linear storage growth if authorized set owners wish to match a subset of the attribute values of an active entity's information. This *partial matching* property is desirable in many situations. For example, customer database A is an authorized owner of some customers' name, credit card number, and mailing address and customer database B is an authorized owner of the same customers' name, credit card number and email addresses. Suppose A and B wish to find out their common costumers by intersecting their respective set of credit card numbers. This cannot be done efficiently with our proposed DOC since A's (resp. B's) customers need to generate one DOC for their names, credit card numbers and mailing addresses (resp. email addresses) respectively. When a database has various information about a customer, the storage overhead can be quite high. In this section, we describe a *Homomorphic Data Ownership Certificate* scheme that allows a customer to generate *one* DOC for all of his or her information submitted to a database and still enables the databases to intersect certain attributes of customer information.

The semantics for a homomorphic data ownership certificate call for a malleable DOC scheme. Given a DOC C_S^O for S from an active entity E, we would like the set owner to generate a valid $\bar{C}_S^O$ for S' where $S' \subset S$ without the help of E.

Homomorphic signatures have the right property we are looking for. Let $\odot$ be a generic binary operator. Intuitively, a homomorphic signature scheme allows anyone to compute a new signature $\mathsf{Sig}(x \odot y)$ given the signatures $\mathsf{Sig}(x)$ and $\mathsf{Sig}(y)$ without the knowledge of the secret key. Johnson et. al introduced basic definitions of security for homomorphic signature systems and proposed several schemes that are homomorphic with respect to useful binary operators [Johnson et al., 2002].

We are interested in the set-homomorphic signature scheme proposed in [Johnson et al., 2002] that supports both union and subset operations. More precisely, the scheme allows anyone to compute $\mathsf{Sig}(S_1 \cup S_2)$ and $\mathsf{Sig}(S')$ where $S' \subseteq S_1$ if he possesses S_1, S_2, $\mathsf{Sig}(S_1)$ and $\mathsf{Sig}(S_2)$.

We now describe our construction for a Homomorphic Data Ownership Certificate (HDOC) scheme. We need to modify both the common nonce and the data ownership component to use the homomorphic signatures. Let S be a set of strings, E the active entity that originates S, and sk_S the signing key exclusively used for S. When E submits its information $S' \subseteq S$ to database A, it issues A an HDOC $H_S^A = \langle pk_S, \mathsf{Sig}_{sk_S}(S'), \mathsf{Sig}_{sk_S}(S' \cup A)\rangle$.

Computing intersection on data with HDOC is straight forward. Suppose databases A and B wish to compute intersection on their customers' credit card number. Then for each customer c_i's HDOC components $\mathsf{Sig}_{sk_S}(S_{c_i})$ and $\mathsf{Sig}_{sk_S}(S_{c_i} \cup A)$, database A computes $\mathsf{Sig}(S'_{c_i})$ and $\mathsf{Sig}(S'_{c_i} \cup A)$ where $S'_{c_i} = \{c_i's \text{ credit card } \#\}$. B does similar computations. Now A and B may run any matching protocol as described in Section 6.2 using the recomputed HDOC.

7. Security Analysis

Recall that we consider four threat models in our paper: the malicious model with a large or small domain, and the semi-honest model with a large or small domain.

We have also identified three goals that a private matching protocol can satisfy: strong/weak, unspoofable/spoofable, and symmetric/asymmetric. In this section, we analyze the effectiveness of the three private matching protocols with respect to each of the threat models and determine what security goals each protocol achieves.

In this section, We analyze the fulfillment of the security goals of the TTPP , HP , AgES , CTTPP , CHP , and CAgES protocols in the four threat models. We summarize the results in Figure 3.2(a) through Figure 3.3(b).

7.1 The malicious model with a large domain

We now analyze the fulfillment of the security goals of the TTPP, HP, AgES, CTTPP, CHP, and CAgES protocols in the malicious model with a large domain. All the unmodified protocols are unspoofable in the absence of collude-then-spoof attacks. Although a large domain makes it difficult for an adversary to guess an element in the other party's set, the adversary can include values obtained from another database in the query to increase the probability of success.

7.1.1 Trusted Third Party Protocol. The Trusted Third Party Protocol (TTPP) is a spoofable, strong and either symmetric or asymmetric matching protocol. TTPP is strong because both parties learn only P and nothing else in a symmetric setting; in an asymmetric setting, one party learns P and the other party learns nothing. TTPP is always strong for this reason in all four threat models. TTPP can be either symmetric or asymmetric depending on whether the sends query results to one or both parties.

Technique		Unspoofable	Strong	Symmetric
TTPP	Sym		X	X
	Asym		X	
HP			(*)	
AgES			X	
CTTPP	Sym	X	X	X
	Asym	X	X	
CHP		$X^{(1)}$	(*)	
CAgES		$X^{(1)}$	X	

(a) Malicious model with a large domain

Technique		Unspoofable	Strong	Symmetric
TTPP	Sym		X	X
	Asym		X	
HP				
AgES			X	
CTTPP	Sym	X	X	X
	Asym	X	X	
CHP		$X^{(1)}$	(*)	
CAgES		$X^{(1)}$	X	

(b) Malicious model with a small domain

Figure 3.2. Security goals satisfied by the protocols in the malicious model. (*): Note that for these examples, we do not have a strong protocol. However, we do have a collusion-free strong protocol which is strong in the absence of colluding attacks . $X^{(1)}$ denotes a protocol is unspoofable in the absence of colluding adversaries.

Technique		Unspoofable	Strong	Symmetric
TTPP	Sym	X	X	X
	Asym	X	X	
HP		X	(*)	
AgES		X	X	
CTTPP	Sym	X	X	X
	Asym	X	X	
CHP		X	(*)	
CAgES		X	X	

(a) Semi-honest model with a large domain

Technique		Unspoofable	Strong	Symmetric
TTPP	Sym	X	X	X
	Asym	X	X	
HP		X		
AgES		X	X	
CTTPP	Sym	X	X	X
	Asym	X	X	
CHP		X	(*)	
CAgES		X	X	

(b) Semi-honest model with a small domain

Figure 3.3. Security goals satisfied by the protocols in the semi-honest model. (*): Note that for these examples, we do not have a strong protocol. However, we do have a collusion-free strong protocol which is strong in the absence of colluding attacks . $X^{(1)}$ denotes a protocol is unspoofable in the absence of colluding adversaries.

7.1.2 Hash Protocol. The Hash Protocol is spoofable, collusion-free strong, and asymmetric. It is strong in the absence of colluding attacks ; since the domain is large, it is difficult for the recipient of a hashed set to guess an element that is actually in the other party's set. However, it is easier for the recipient of a hashed set to learn whether an element is in the other party's set if the recipient uses values obtained from another database in the matching. The Hash Protocol is asymmetric since the party that receives the result first may or may not send the (correct) result to the other party.

7.1.3 AgES protocol. The AgES protocol is spoofable, strong, and asymmetric. The AgES is strong because no attacker may learn any additional information besides the query result and the size of the other party's set. It is asymmetric since the party that receives the result first may or may not send the (correct) result to the other party.

7.1.4 Certified matching protocols. CTTPP is unspoofable. If one of the parties spoofs some element d, the trusted third party can detect it by checking the ownership attestation component as described in Section 6.2.1.

Both CHP and CAgES are unspoofable in the absence of colluding adversaries. The common nonces in a DOC prevent a party from guessing the correct nonce associated with certain data and thus prevent guess-then-spoof attacks.

When colluding parties exist, CHP and CAgES are spoofable. Assume Alice is an authorized owner of some information d and Charlie is not. Alice colludes with Charlie and gives data d and the associated DOC to Charlie. When Bob sends his data set to Charlie in a CHP execution, Charlie can learn whether Bob has d by hashing the nonce n_d associated with d and checking if it is in Bob's set. There is a non-negligible probability that n_d is in Bob's set. This matching result violates the definition for unspoofable. Similarly, in a CAgES protocol execution, Charlie may encrypt the nonce n_d and send it to Bob. Charlie will discover whether Bob has d when Bob honestly responds to the query.

On the other hand, if Charlie and Bob switch roles in the CHP and CAgES protocol executions, Charlie cannot prove to Bob that he has d since he does not have a valid ownership attestation component for d.

7.2 The malicious model with a small domain

With a small domain, a malicious adversary can guess an element of the other party's set with non-negligible probability. An adversary can then launch a spoofing attack and learn elements of the other party's set not contained in its own with non-negligible probability. Therefore, without modification, all three protocols are spoofable in the malicious model with a small domain.

7.2.1 Trusted Third Party Protocol. The trusted third party is spoofable, strong, and either symmetric or asymmetric. The analysis is similar to that of the malicious model with a large domain presented in Section 7.1.1.

7.2.2 Hash Protocol. The hash protocol is spoofable, weak, and asymmetric. It is weak because a malicious party may launch a guess-then-spoof attack and succeed in learning the entire set of the other party with high probability. The analysis for asymmetry is similar to that of the hash protocol for the malicous model with a large domain presented in Section 7.1.2.

7.2.3 AgES protocol. The AgES protocol is spoofable, strong, and asymmetric. The AgES is spoofable because although the encryption scrambles the data, it cannot prevent spoofing attacks. The analysis for AgES being strong is similar to that of the malicious model with a large domain in Section 7.1.3. The analysis for asymmetry is similar to that of a large domain presented in Section 7.1.3.

7.2.4 Certified matching protocols. By combining the DOC with TTPP, HP, and AgES, we obtain protocols that satisfy the same security properties in the malicious model with a small domain as the corresponding certified protocols in the malicious domain with a large domain. In particular, by adding the DOC component, we enable the protocol to detect spoofed queries in the absence of colluding attacks .

7.3 The semi-honest model with a large domain

All three protocols are trivially unspoofable in a semi-honest model since parties do not cheat in a semi-honest model. For the strong/weak dimension, each protocol satisfies the same security goal as the corresponding protocol in a malicious model with a large domain in Section 7.1. The TTPP is can be either symmetric or asymmetric depending on whether the trusted party sends the result to one or both parties. The HP and AgES can also be either symmetric or asymmetric depending on whether the protocol prescribes the party which receives the result first sends it to the other party.

The TTPP is unspoofable, strong, and symmetric/asymmetric. The analysis of TTPP being strong is similar to that of a large domain presented in Section 7.1.1.

The AgES is an unspoofable, strong, and symmetric/asymmetric and protocol. The analysis of AgES beging strong is similar to that of a malicious model with a large domain in Section 7.1.3.

7.3.1 Certified matching protocols. All unmodified protocols are unspoofable in the semi-honest model. The DOC mechanism is not applicable in the semi-honest model with a large domain and this becomes clear in Section 7.4.

7.4 The semi-honest model with a small domain

The analysis for the semi-honest model with a small domain is similar to that of the semi-honest model with a large domain. The only difference is that the HP is collusion-free strong in the large domain and weak in the small domain and by combining the DOC with the HP, we obtain a protocol that is collusion-free strong in the semi-honest model with a small domain.

Protocol	Cost	Complexity
TTPP	$q \log q + b \log q$	$O(b \log b)$
HP	$C_h(q+b) + b \log b + q \log b$	$O(b \log b)$
AgES	$(C_h + 2C_e)(q+b) + 2b \log b + 3q \log q$	$O(C_e b)$
CTTPP	$q \log q + b \log q + 2C_x r$	$O(C_x r)$
CHP	$C_h(q+b) + b \log b + q \log b + C_x r$	$O(C_x r)$

(a) Computational cost

Protocol	Cost	Complexity
Asymmetric TTPP	$(q+b+r) \cdot n$	$O(bn)$
Symmetric TTPP	$(q+b+2r) \cdot n$	$O(bn)$
HP	$b \cdot l$	$O(bl)$
AgES	$(2q+b) \cdot m$	$O(bm)$
Asymmetric CTTPP	$(q+b+r) \cdot n + (q+b) \cdot k$	$O(bn)$
Symmetric CTTPP	$(q+b+r) \cdot n + (q+b) \cdot k$	$O(bn)$
CHP	$(l+k) \cdot b$	$O(bk)$

(b) Communication cost

Figure 3.4. Cost analysis

8. Cost Analysis

In this section, we use the following notations. Alice poses a query Q to Bob who has a set B. Let $P = Q \cap B$ be the query result. Let $q = |Q|$, $b = |B|$, and $p = |P|$. Let C_h be the cost of hashing and C_x be the cost of running the public key verification algorithm $\text{VERIFY}(\cdot, \cdot, \cdot)$. Let j be the length of a public key, k be the length of the ownership attestation component, l be the length of the output of $h(\cdot)$, m be the length of each encrypted code word in the range of $\mathcal{F}$, and n be the length of each element; all quantities are in bits. We assume that the set Q is larger than the set B, i.e. $b < q$, and we assume that $l \leq k + j \leq n$.

We present the computational and communication cost in Figure 3.4(a) and Figure 3.4(b) respectively.

The computational costs of the trusted third party and hashing protocols are dominated by the cost of sorting the list. For the AgES and certified protocols, the computation cost is dominated by the encryption/decryption and public key signature verification respectively. Further details can be found in Figure 3.

As we may see from Figure 3.4(b), the communication cost for any proposed protocol is linear in the size of the sets being sent. This linear communication cost is the lower bound of any set intersection protocols which compute exact matching [Kalyanasundaram and Schnitger, 1992].

9. Related Work

Private Information Retrieval (PIR) schemes allow a user to retrieve the i-th bit of an n-bit database without revealing i to the database [Beimel and Ishai, 2001, Cachin et al., 1999, Chor et al., 1995]. These schemes guarantee user privacy. Gertner et al. introduce Symmetrically-Private Information Retrieval (SPIR) where the privacy of the data, as well as the privacy of the user is guaranteed [Gertner et al., 1998]. In every invocation of a SPIR protocol, the user learns only a single bit of the n-bit database, and no other information about the data. Practical solutions are difficult to find since the PIR literature typically aims for very strong information-theoretic security bounds.

There has been recent work on searching encrypted data [Boneh and Franklin, 2004, Waters et al., 2004] inspired by Song, Wagner, and Perrig's original paper describing practical techniques for searching encrypted data [Song et al., 2000]. Song et al. proposed a cryptographic scheme to allow a party C to encrypt and store data on an untrusted remote server R. R can execute encrypted queries issued by C and return encrypted results to C.

10. Future Work

This paper explores some issues associated with private matching. But many areas remain to be explored. Here, we list a few particularly interesting challenges:

- In this paper, we examined two party protocols. What are the issues that arise with more complicated protocols with more than two parties?

- There is a basic asymmetry that arises between two parties where one party knows significantly more than a second party. Parties that control large sets may be able to extract significantly more interesting information than parties that control small sets. There may be instances where parties controling small sets can detect and reject these queries.

- Here, we only consider examples of matching elements from two sets. More interesting and more far-ranging examples are possible. For instance, this paper considered *listing queries* — we actually listed all the elements held in common between two sets. We can consider a broader range of *functional queries* which return a function calculated over the intersection of two sets. While a broad literature in statistical databases exists, the question of functional operations is a more general notion that deserves further attention.

- There is an interesting connection between our spoofing discussion and the database literature on updates through views. The view update literature provides (constrained) solutions for the following: given a query on relation instances R and S resulting in a set P, what changes to R and S could produce some new answer P'? The reasoning used to address that problem is not unlike the reasoning used to learn information via spoofing: by substituting R' for R and observing the query result P', what can be learned about S? The literature on updates through views is constrained because it seeks scenarios where there is a unique modification to R, S that can produce P'. By contrast, much can be learned in adversarial privacy attacks by inferring a non-unique set of *possible* values for S.

- In large distributed systems, it may be desirable to have a set of peer systems store information in a variety of locations. In this broader distributed system, can we still guarantee privacy properties.

- In our list of attacks in Section 3.1, we discussed a hiding attack where a database owner pretends certain values don't occur in its database. Can we provide effective defenses against hiding attacks?

Notes

1. In the introduction, we argued that semi-honest protocols were unrealistic in many situations. However, for completeness we will consider them here.

References

[Agrawal et al., 2003] Agrawal, R., Evfimievski, A., and Srikant, R. (2003). Information sharing across private databases. In *Proceedings of the 2003 ACM SIGMOD Int'l Conf. on Management of Data, San Diego, CA.*

[Agrawal et al., 2002] Agrawal, Rakesh, Kiernan, Jerry, Srikant, Ramakrishnan, and Xu, Yirong (2002). Hippocratic databases. In *28th Int'l Conf. on Very Large Databases (VLDB), Hong Kong.*

[Beimel and Ishai, 2001] Beimel, Amos and Ishai, Yuval (2001). Information-theoretic private information retrieval: A unified construction. *Lecture Notes in Computer Science*, 2076.

[Bellare et al., 1996] Bellare, M., Canetti, Ran, and Krawczyk, Hugo (1996). Keying hash functions for message authentication. In *Proceedings of the 16th Annual International Cryptology Conference on Advances in Cryptology table of contents*, pages 1–15. Lecture Notes In Computer Science archive.

[Boneh and Franklin, 2004] Boneh, D. and Franklin, M. (2004). Public key encryption with keyword search. In *Eurocrypt 2004, LNCS 3027*, pages 56–73.

[Cachin et al., 1999] Cachin, Christian, Micali, Silvio, and Stadler, Markus (1999). Computationally private information retrieval with polylogarithmic communication. *Lecture Notes in Computer Science*, 1592.

[Canetti, 1996] Canetti, Ran (1996). *Studies in Secure Multiparty Computation and Applications*. PhD thesis, The Weizmann Institute of Science.

[Chor et al., 1995] Chor, Benny, Goldreich, Oded, Kushilevitz, Eyal, and Sudan, Madhu (1995). Private information retrieval. In *IEEE Symposium on Foundations of Computer Science*, pages 41–50.

[Gertner et al., 1998] Gertner, Yael, Ishai, Yuval, Kushilevitz, Eyal, and Malkin, Tal (1998). Protecting data privacy in private information retrieval schemes. In *The Thirtieth Annual ACM Symposium on Theory of Computing*, pages 151–160.

[Goldreich et al., 1987] Goldreich, O., Micali, S., and Wigderson, A. (1987). How to play any mental game. In *Proc. of 19th STOC*, pages 218–229.

[Goldreich, 2002] Goldreich, Oded (2002). Secure multi-party computation. Final (incomplete) draft, version 1.4.

[Huberman et al., 1999] Huberman, B. A., Franklin, M., and Hogg, T. (1999). Enhancing privacy and trust in electronic communities. In *ACM Conference on Electronic Commerce*, pages 78–86.

[Johnson et al., 2002] Johnson, Robert, Molnar, David, Song, Dawn Xiaodong, and Wagner, David (2002). Homomorphic signature schemes. In *CT-RSA*, pages 244–262.

[Kalyanasundaram and Schnitger, 1992] Kalyanasundaram, B. and Schnitger, G. (1992). The probabilistic communication complexicty of set intersection. SIAM J. Discrete Mathematics.

[Menezes et al., 1996] Menezes, A., van Oorschot, P., and Vanstone, S (1996). Handbook of applied cryptography. CRC Press.

[Parliament, 1995] Parliament, European (1995). Directive 95/46/EC of the European Parliament and of the Council of 24 October 1995 on the protection of individuals with regard to the processing of personal data and on the free movement of such data.

[Song et al., 2000] Song, Dawn Xiaodong, Wagner, David, and Perrig, Adrian (2000). Practical techniques for searches on encrypted data. In *IEEE Symposium on Security and Privacy*, pages 44–55.

[Waters et al., 2004] Waters, Brent R., Balfanz, Dirk, Durfee, Glenn, and Smetters, D.K. (2004). Building an encrypted and searchable audit log. In *The 11th Annual Network and Distributed System Security Symposium*.

[Yao., 1986] Yao., A.C. (1986). How to generate and exchange secrets. In *In Proceedings of the 27th IEEE Symposium on Foundations of Computer Science*, pages 162–167.

Chapter 4

AUTHENTICATION PROTOCOL ANALYSIS

Jonathan Millen
SRI International
Menlo Park, CA, USA
millen@csl.sri.com

Abstract We take a closer look at some of the limitations of current analysis approaches, and mention some work and open problems related to expanding their scope.

Keywords: cryptographic authentication protocol analysis, decidable network security, constraint solver

1. Introduction

Security analysis of cryptographic protocols is getting to be an old, classical subject. The mathematical approach to protocol analysis, as opposed to cryptosystem analysis, goes back to [Dolev and Yao, 1983]. The essential ideas in that paper were in abstract algebraic treatment of encryption, and the security threat in the form of an active entity in the network that could intercept, redirect, and construct messages. This interest in formal protocol analysis was motivated by the observation that security analysis was too subtle for reliable informal analysis. The first influential published example of the kind of vulnerability that escaped informal analysis was the replay attack in [Denning and Sacco, 1981] on the symmetric-key protocol in [Needham and Schroeder, 1978]. This showed that the security of a protocol could be compromised without breaking the encryption algorithm.

The Dolev-Yao results were still not powerful enough to apply to the Needham-Schroeder protocols, so other methods such as model checking and inductive verification on state-transition models were developed. Current research is aimed at making these methods more efficient and expanding their scope. The efficiency of some tools is remarkable; current model checking tools can analyze typical protocols in seconds or fractions of seconds.

The objective of this note is to take a closer look at some of the limitations of current analysis approaches, and mention some work and open problems related to expanding their scope. We consider both limitations in the protocol modeling approach and limitations in the analysis approach. A protocol modeling approach may be inadequately expressive in several respects, such as:

- vocabulary of cryptographic and computational operators,
- representation of recursive or group protocol structure, or
- security properties.

Limitations on the validity of conclusions obtained from analysis using formal methods may arise partly from:

- incomplete axiomatization of individual operators,
- assumptions about type checking or the lack of it,
- weakness in the attacker model, or
- resource limitations in the vulnerability search method.

Some of the more interesting problems and results in these areas are discussed below.

2. Modeling Computational Operations

Messages are typically represented as algebraic terms over one or more sorts of primitive data such as addresses and keys. Operators include at least encryption and pairing, and a hash function. Public-key and symmetric encryption are distinguishable axiomatically, and one expects them both to be represented. Differences between cryptosystems with the same axiomatic properties, such as DES and AES, do not show up in the algebraic axioms, so different block ciphers need not be distinguished. This much is expected.

If all we know about a encryption is represented with abstract rules like $\mathrm{d}(\mathrm{e}(X, K), K) = X$, it is obvious that many properties of the operator and many possible attacks might have been lost. The virtue of abstract analysis at this level is that many possible attacks can still be found, and, with a few exceptions, an attack that works at this level can be made to work on the real protocol. Thus, the analysis approach is conservative, from a security point of view. Furthermore, if the analysis approach finds an unimplementable attack, it is usually easy to check that it will not work. With a carefully chosen implementation, it may even be the case that there are no other attacks. A completeness statement like this is the goal of some work, discussed below, regarding the computational soundness of formal encryption.

Even at this extreme level of abstraction, there are design choices. For example, the ability to decrypt a (symmetrically) encrypted data item may be represented either as a reduction rule $\mathrm{d}(\mathrm{e}(X, K), K) = X$ or as an attacker derivation rule $\mathrm{e}(X, K), K \vdash X$. The derivation rule approach is used when it is desired to use a free algebra for message terms (so that "d" need not be explicit). The use of a free algebra has some advantages for justifying and implementing analysis approaches, but some attacks might be lost, as discussed in [Millen, 2003]. Even if "d" is not required explicitly to state the protocol, this formalism cannot represent the ability of an attacker to decrypt data that has not already been encrypted.

For example, consider a (very artificial) protocol in which one party sends a secret s encrypted with a secret key k, as $\mathrm{e}(s, k)$. Assume that a second party replies to any message of the form $\mathrm{e}(\mathrm{e}(X, c), k)$ with X, where X is a pattern variable and c is a key known to the attacker. This protocol compromises s if we can write the equation $s = \mathrm{e}(X, c)$, since X is then sent by the second party and the attacker knows c. However, that equation cannot be satisfied in a free algebra, where terms are equal only if identical. Thus, analysis based on a free algebra will overlook the compromise.

It is shown in [Millen, 2003] that the culprit is $\mathrm{e}(X, c)$, in which an unknown received item X is encrypted alone. If such items are always encrypted in some context, such as $\mathrm{e}([X, a], c)$, no attacks are overlooked. This was shown first for symmetric encryption , and the result has been extended to public key encryption by [Lynch and Meadows, 2004].

Other operations, such as exclusive-or (bitwise binary addition), have properties that are not representable with a free algebra. As a commutative operation, exclusive-or shares some of the difficulty of exponentiation as used in Diffie-Hellman , which is discussed below.

Another issue with abstract models is the handling of concatenation. The simplest models allow only the formation of pairs $[X, Y]$. If a message has three fields, it might be written either as $[[X, Y], Z]$ or $[X, [Y, Z]]$, and one might ask whether these two forms are distinguishable; that is, is pairing considered to be associative? In many models it is, but in a protocol that implements pairing as concatenation of bit strings, it is not. Thus, an attack that rests on indistinguishability of the two forms will be overlooked in the abstract analysis.

One might also ask whether longer lists are distinguishable from nested pairs. Is $[X, Y]$ distinguishable from a triple $[X, U, V]$? What if $Y = [U, V]$? There is an interesting (but academic) attack on a modified version of the Needham-Schroeder public-key handshake that plays on this confusion. Lowe's version of the protocol, generally regarded as secure, is:

$$A \rightarrow B : [N_a, A]\text{pk}(B)$$
$$B \rightarrow A : [N_a, N_b, B]\text{pk}(A)$$
$$A \rightarrow B : [N_b]\text{pk}(B)$$

The important modification is the addition of B in the second message. If we change the first message to $[A, N_a]\text{pk}(B)$, an apparently trivial change, an attack is possible, given in [Millen and Shmatikov, 2001]. The intruder creates a message $[A, E]\text{pk}(B)$ to get the reply $[E, N_b, B]\text{pk}(A)$, and sends this to A, hoping that A will act as a responder, see this message as $[E, [N_b, B]]pk(A)$, and reply with $[[N_b, B], N_a, A]\text{pk}(E)$ so that E can read N_b. This formal attack is likely to fail in reality for several reasons, but it is interesting because it shows that

- some formal attacks might not be exploitable,
- the distinction between $[X, Y, Z]$ and $[X, [Y, Z]]$ is worth thinking about, and
- protocol analyzers can find surprising attacks.

This example also raises questions about what kind of type checking is done in real protocol implementations. Cryptographic APIs may be very sophisticated about the way encryption is implemented, but message construction is usually a simple processing of bit strings. Separation of a message into fields is accomplished by a combination of counting off the bit-length of fields of an expected size, plus some checking of the contents of fields with expected values. Abstract protocol models usually do not represent this procedure accurately. Many protocol models assume that strong type checking is done, typically because the model is stated and analyzed in some formal language or environment where type assignment is mandatory. Protocol models with free algebras assume that one can distinguish data with different term representations, such as a key constant K from an encryption $\text{e}(X, K)$, even though the encrypted result might be the same length, and might even be used as a key. With regard to type checking, is it worth noting the result in [Heather, et al., 2003], which shows how type flaw attacks can be prevented by adding encryption-protected tags to data fields.

For models of symmetric encryption , it is a desirable feature to allow non-atomic or computed keys. In SSL , the protocol used in standard Web browsers, for example, a "master secret " is computed by hashing a "pre-master secret" and some random data. Similar constructions are used in many Internet protocols to construct symmetric keys or nonces. A model that does not allow computed keys is unrealistically limited. On the other hand, a model that allows any computational result to be a key is unrealistically permissive. The question of what computed data can be used as a key (or an address, or other message field) is related to questions about type checking.

3. Diffie-Hellman and Group Protocols

The seminal key-agreement protocol is the Diffie-Hellman protocol, in which two parties have private secrets x and y, and they exchange exponentials $a^x \mod p$ and $a^y \mod p$. Then they can both compute a common key $a^{xy} \mod p = a^{yx} \mod p$ without disclosing the secrets to each other or to anyone else. Can this kind of protocol be axiomatized abstractly without incorporating a complete theory of finite-field arithmetic? As a reasonable first step, one could introduce two abstract operations $\exp(x, y)$ representing $x^y \mod p$, and $g(x)$ representing $a^x \mod p$, with the relation $\exp(g(x), y) = \exp(g(y), x)$. (Something like this was suggested in [Blanchet, 2001].) This representation is good enough to show how Diffie-Hellman works in some protocols, but it is limited. It cannot show that $\exp(\exp(g(x), y), z) = \exp(\exp(g(y), z), x)$, for example, and some protocols and some attacks might depend on that.

A few papers have begun to investigate how Diffie-Hellman could be handled in an abstract but adequate way for protocol analysis ([Meadows and Narendran, 2002, Microsoft, 2003, Chevalier, et al., 2003]). Part of the approach is to solve unification problems involving nested exponentials, like a^{xyz}, by reducing them to unification problems for products of exponents xyz. The product is understood as an Abelian group operation. There are some decidability results, but there are not yet any efficient algorithms for performing the analysis of protocols with exponentiation.

The Diffie-Hellman idea can be extended to key agreement protocols with three or more parties in a natural way. The result is group Diffie-Hellman. A good axiomatization of Diffie-Hellman, together with the associated unification and other techniques, can take care of group Diffie-Hellman as well. However, group protocols are challenging to specify and analyze for additional reasons. A group protocol, by definition, involves a variable and dynamically changing number of participants. Multicast or broadcast of messages is usually required. Furthermore, messages may include lists, the lengths of which vary according to the current number of members participating in the group. Finally, new kinds of operations, like threshold encryption , are useful for groups. There have been some experiments with automated analysis of group protocols, but there are still no general tools available for them. There is an interesting example of a protocol vulnerability in a protocol using group Diffie-Hellman, in [Pereira and Quisquater, 2001]. That example may very well turn into the "Needham-Schroeder"-like standard example for group protocol analysis methods to come.

4. Deeper Models of Encryption

Security protocol analysis can be conducted using computational methods with more detailed models of cryptographic algorithms. These methods examine complexity and probabilistic issues. While encryption is all-or-none in the ideal models used in more abstract analysis, computational methods can detect attacks that cause partial information to be "leaked." The computational approach can explain attacks like the Bleichenbacher attack , which takes advantage of malleability in basic RSA to extract a plaintext message, a fraction of a bit at a time, over many sessions of the PKCS #1 protocol (see [Bleichenbacher, 1998]).

Malleability is the ability to modify the decrypted plaintext in a predictable way while it is still encrypted. In the case of PKCS #1, the vulnerability arises from the homomorphic property of RSA : $(xy)^e = x^e y^e$. Understanding exponentials, as for Diffie-Hellman, is not enough to uncover the attack; probabilistic arguments are needed also.

With certain conditions on the cryptographic algorithm, one can state that it is "ideal" in the sense that it is indistinguishable, according to a probabilistic complexity model, to its abstractly axiomatized representation. The first result of this kind was in [Abadi and Rogaway, 2002]. With ideal cryptographic primitives, security conclusions from the abstract analysis should carry over to the implementation. One of the first papers to study the relationship of computational to formal models is [Abadi and Rogaway, 2002]. This work looked at message terms in isolation. The next step was to extend that result to a sequence of messages in a protocol. This is the approach taken by [Backes, et al., 2003]. They showed how to implement cryptographic primitives in a protected way, through a controlled interface designed to remove all traces of malleability. Their abstract interface is different from the usual abstract model, so that it will be a challenge for formal methods to adapt to the differences.

5. Decidable Formal Methods

Despite the simplicity of models used in abstract methods, security analysis is undecidable in an environment with active attacks and an unbounded number of protocol processes. This has been proved with several different models in different ways from the the first undecidability proof in 1983 by [Even and Goldreich, 1983], and more recently, for example, in [Durgin, et al., 1999]. Undecidability arises partly from data types that are viewed abstractly as unbounded, such as the key space, and the unbounded number of concurrent protocol sessions. The fact that these parameters are bounded in reality doesn't help much; the undecidability result reflects real difficulties in the complexity of analysis. It has been shown that if the number of concurrent *honest* protocol processes is unbounded, then secrecy analysis is decidable, but it is still NP-complete (see [Rusinowitch and Turuani, 2001]).

When the number of sessions is bounded, the analysis approach still has to deal with the problem of unbounded data in messages and the unbounded activity of the intruder. [Huima, 1999] suggested using a symbolic approach, in which message subfields were represented by variables that were not instantiated unless necessary. This permits the symbolic message space to be bounded for purposes of analysis. His approach was not fully presented, but the ideas were developed further by several researchers.

The SRI constraint solver presented in [Millen and Shmatikov, 2001] implements the symbolic approach in an unusual but efficient and understandable way. A protocol is converted to a set of algebraic term closure constraints. The constraints have a solution if and only if a security vulnerability exists. Term closure constraints require new techniques to solve. They are not, for example, just set constraints, which are not adequate to describe cryptographic protocols without some extensions, as discussed by [Comon, et al., 2001].

An analysis problem is presented as a *semibundle* , which is a finite collection of symbolic *strands* representing protocol processes in the sense of [Thayer, et al., 1999]. A strand is a sequence of nodes representing message send and receive events. Each role in a protocol can be represented as a schema, that is instantiated to form strands for processes in that role. There may be any number of strands in a semibundle, since each role in the protocol may be instantiated any number of times. A semibundle that represents a realizable partially ordered message history is a *bundle*. In a bundle, received messages can be identified with sent messages. One must keep in mind that a sent message may have been computed by an attacker from previously sent messages. The standard strand space model uses primitive "penetrator" strands for allowable attacker computations. A tool called Athena was the first to treat the analysis problem as a search for a way to complete a semibundle to a bundle, in [Song, 1999].

The constraint solver does not use penetrator strands. Instead, it expresses the computation requirement directly by enumerating the possible sequential (send and receive) event histories. Then, for each received message in a given history, the computation condition is expressed as a term closure constraint. The closure of a set of (sent message) terms is the collection of messages that can be computed from it; this closure must contain the next received message. Security violations are expressed with special "test" strands. Security is violated for a semibundle if there is a history containing a test strand in which the constraints for all received messages are satisfied.

It is not necessary to generate all possible histories. The original constraint solver had a "send-first" optimization that eliminated some. Others can be eliminated by solving constraint sets incrementally, to discard whole subtrees of unsolvable cases (see [Corin and Etalle, 2002]). This optimization has been incorporated into the Prolog constraint solver on the SRI CSL web site. Recently, a group at ETH Zurich found another significant optimization called "constraint differentiation;" see [Basin, et al., 2003].

6. Future Directions

Several challenges for new security protocol analysis tools have been mentioned:

- Recognize the limitations of the free algebra model for messages.
- Allow for computed symmetric keys.
- Represent concatenations appropriately.
- Make realistic type-checking assumptions.
- Handle Diffie-Hellman in a general way.
- Represent and analyze group protocols.
- Take advantage of results based on computational methods.

Despite advances in computational methods, formal methods remain necessary for protocol analysis. Within their own domain, modern analysis tools must deal in some constructive and useful way with the known sources of undecidability and complexity in protocol analysis, such as unbounded data types, unbounded intruder activity, and an unbounded number of protocol processes. At the same time, they must be efficient when applied to suitably small examples. The constraint solver is a tool that is up to date with respect to decidability results, straightforward enough conceptually to inspire confidence in its soundness, and flexible enough to encourage further development.

References

[Abadi and Rogaway, 2002] M. Abadi and P. Rogaway (2003). Reconciling two views of cryptography. *J. Cryptology* (15)2, pages 103–127.

[Burrows, Abadi, and Needham, 1990] M. Burrows, M. Abadi and R. Needham (1990). A logic of authentication. *ACM Transactions on Computer Systems* 8(1), pages 18–36.

[Blanchet, 2001] B. Blanchet (2001) An efficient cryptographic protocol verifier based on Prolog rules. *14th IEEE Computer Security Foundations Workshop*, pages 82–96.

[Bleichenbacher, 1998] D. Bleichenbacher (1998). Chosen ciphertext attacks against protocols based on the RSA encryption standard PKCS #1. In

Advances in Cryptology - CRYPTO '98, volume 1462 of *LNCS*, pages 1–12. Springer.

[Basin, et al., 2003] D. Basin, S. Moedersheim, and L. Vigano (2003). Constraint differentiation: A new reduction technique for constraint-based analysis of security protocols. In *ACM Conference on Computer and Communication Security*. ACM SIGSAC.

[Backes, et al., 2003] M. Backes, B. Pfitzmann, and M. Waidner (2003). A composable cryptographic library with nested operations. In *ACM Conference on Computer and Communications Security*. ACM SIGSAC, 2003.

[Comon, et al., 2001] Hubert Comon, Véronique Cortier, and John Mitchell (2001). Tree automata with one memory, set constraints, and ping-pong protocols. *Lecture Notes in Computer Science*, 2076, pages 682–693.

[Corin and Etalle, 2002] R. Corin and S. Etalle (2002). An improved constraint-based system for the verification of security protocols. In *9th Int. Static Analysis Symp. (SAS)*, volume LNCS 2477, pages 326–341. Springer-Verlag.

[Chevalier, et al., 2003] Y. Chevalier, R. Kuesters, M. Rusinowitch and M. Turuani (2003). Deciding the security of protocols with Diffie-Hellman exponentiation and products in exponents. IFI-Report 0305, CAU Kiel.

[Durgin, et al., 1999] N. Durgin, P. Lincoln, J. Mitchell, and A. Scedrov (1999). Undecidability of bounded security protocols. *Formal Methods and Security Protocols*, FLOC 99.

[Denning and Sacco, 1981] D. Denning and G. Sacco (1981). Timestamps in key distribution protocols. *Communications of the ACM* 24(8).

[Dolev and Yao, 1983] D. Dolev and A. Yao (1983). On the security of public key protocols. *IEEE Transactions on Information Theory*, IT-29. pages 198–208, Also STAN-CS-81-854, May 1981, Stanford U.

[Even and Goldreich, 1983] S. Even and O. Goldreich (1983). On the security of multi-party ping-pong protocols. *24th IEEE Symposium on Foundations of Computer Science*.

[Heather, et al., 2003] J. Heather, G. Lowe and S. Schneider (2003). How to prevent type flaw attacks on security protocols. *13th IEEE Computer Security Foundations Workshop*, pages 255–268.

[Huima, 1999] A. Huima (1999). Efficient infinite-state analysis of security protocols. In *Workshop on Formal Methods and Security Protocols*, FLOC.

[Lowe, 1996] G. Lowe (1996). Breaking and fixing the Needham-Schroeder public-key protocol using FDR. In *Proceedings of TACAS*, volume 1055 of *Lecture Notes in Computer Science*, pages 147–166. Springer-Verlag.

[Lynch and Meadows, 2004] C. Lynch and C. Meadows (2004). On the relative soundness of the free algebra model for public key encryption. In *Workshop on Issues in the Theory of Security (WITS)*, IFIP WG 1.7.

[Meadows and Narendran, 2002] C. Meadows and P. Narendran (2002). A unification algorithm for the group Diffie-Hellman protocol. *Workshop on Issues in the Theory of Security (WITS 02)*, pages 1–10.

[Millen, 2003] J. Millen (2003). On the freedom of decryption. *Information Processing Letters*, 86(6): pages 329–333.

[Millen and Shmatikov, 2001] J. Millen and V. Shmatikov (2001). Constraint solving for bounded-process cryptographic protocol analysis. In *8th ACM Conference on Computer and Communication Security*, pages 166–175. ACM SIGSAC, November.

[Millen and Shmatikov, 2003] J. Millen and V. Shmatikov (2003). Symbolic protocol analysis with products and Diffie-Hellman exponentiation. *16th IEEE Computer Security Foundations Workshop*, pages 47–61.

[Needham and Schroeder, 1978] R. Needham and M. Schroeder (1978). Using encryption for authentication in large networks of computers. *Communications of the ACM* (21)12, December, pages 993–998.

[Pereira and Quisquater, 2001] O. Pereira and J. Quisquater (2001). A security analysis of the Cliques protocol suites. *14th IEEE Computer Security Foundations Workshop*, pages 73–81.

[Perrig and Song, 2000] A. Perrig and D. Song (2000). A first step toward the automatic generation of security protocols. *Network and Distributed System Security Symposium*.

[Rusinowitch and Turuani, 2001] M. Rusinowitch and M. Turuani (2001). Protocol insecurity with finite number of sessions is NP-complete. In *14th IEEE Computer Security Foundations Workshop*, pages 174–190. IEEE Computer Society.

[Song, 1999] D. Song (1999). Athena: a new efficient automatic checker for security protocol analysis. In *12th IEEE Computer Security Foundations Workshop*, pages 192–202. IEEE Computer Society.

[Thayer, et al., 1999] J. Thayer, J. Herzog, and J. Guttman (1999). Strand spaces: proving security protocols correct. *Journal of Computer Security*, 7(2/3): pages 191–230.

Chapter 5

SELF-CERTIFIED APPROACH FOR AUTHENTICATED KEY AGREEMENT

Tzong-Chen Wu
Department of Information Management
National Taiwan University of Science and Technology
Taipei, Taiwan
email: tcwu@cs.ntsut.edu.tw

Yen-Ching Lin
Department of Information Management
National Taiwan University of Science and Technology
Taipei, Taiwan
email: D9109101@mail.ntust.edu.tw

Abstract Password-only authenticated key agreement (or PAKA for short) protocols allow communication parties to mutually authenticate with each other and share an authenticated secret key by only using easy-to-remember passwords. In this paper, we present a point-to-point PAKA protocol (or 2-PAKA for short) based on self-certified approach. The proposed 2-PAKA can be easily generalized to a point-to-multipoint PAKA (or n-PAKA for short) that allows n communication parties to achieve mutual authentication and key agreement. The proposed PAKA protocols achieve the properties of perfect forward secrecy and known-key security. Communication messages produced by the proposed PAKA protocols are self-certified, and therefore no trusted servers or public key certificates are required during the key agreement phase. We also discuss some essential but potential attacks on the proposed PAKA protocols, including on-/off-line password guessing, password-compromised impersonation, and unknown key-share.

1. Introduction

Using easy-to-remember passwords for user authentication is widely adopted in contemporary computer systems, because of its ease of use, cost effective, and ease of implementation. In 1992, Bellovin and Merritt [Bellovin and

Merritt, 1992] proposed a family of password-based encrypted key exchange (EKE) protocols, whereby any two communication parties shared with a pre-chosen password in advance can exchange a session key . In 1995, Steiner, Tsudik and Waidner [Steiner *et al.*, 1995] successfully modified the Bellovin-Merritt's protocols into the so-called three-party EKE protocols . Their protocols require a trusted server during the key agreement phase. Any two communication parties (act as clients) can only use their passwords to achieve mutual authentication and key agreement with the assistance of the trusted server. Since then, several password-based three-party EKE protocols are developed [Bellare *et al.*, 2000, Bresson *et al.*, 2000, Lee *et al.*, 1999, Lin *et al.*, 2001, MacKenzie *et al.*, 2000]. We call such type of key agreement protocols as the PAKA protocols for short.

Basically, an authenticated key agreement protocol shall achieve the following security requirements addressed in [Blake-Wilson and Menezes, 1998, Ding and Horster, 1995]:

1 *It is known-key security.* That is, an attacker cannot derive any session keys established between the communication parties from any compromised session key .

2 *It is perfect forward secrecy.* That is, an attacker cannot derive any previously established session keys from a compromised password which is regarded as a long-term private key.

3 *It is resistant to on-/off-line password guessing attacks.* That is, an attacker cannot find out the passwords, even though it is easy-to-remember, from the intercepted messages by using an or by using a password dictionary.

4 *It is resistant to password-compromised impersonation.* Suppose that an attacker compromised a party U_i's password PW_i. Clearly, he can thoroughly impersonate U_i. However, it may be desirable in some circumstances that the attacker cannot impersonate any other parties, say U_j, to U_i using the compromised PW_i.

5 *It is resistant to unknown key-share attacks.* The scenario of launching such attack is as follows: An attacker intercepted the communication messages originated by one party U_i and then replayed, or modified and resent, these messages to the other party U_j. For the success of such attack, U_i ends up believing that he shares a session key with U_j, and although this is in fact the case, U_j mistakenly believes that the session key is instead shared with some other party $U_a \neq U_i$.

In this paper, we first present a point-to-point PAKA protocol (or 2-PAKA for short) based on self-certified approach, and then extend the proposed 2-

PAKA to a point-to-multipoint PAKA (or n-PAKA for short). The proposed PAKA protocols achieve the security requirements described above. Meanwhile, the proposed PAKA protocols require no trusted servers or public key certificates for assistance during the key agreement phase, because the communication messages are self-certified.

2. Proposed 2-PAKA Protocol

The proposed 2-PAKA protocol requires a system authority (SA), whose tasks are to initialize necessary parameters for system setup and to accept user's registration via an interception-resistant channel. Note that SA does not know the registering user's password; instead, SA just knows the protected password blobbed by a public one-way function. The proposed 2-PAKA protocol is divided into three phases: system setup, user registration, and key agreemenet. Details of these phases are described in the following.

System setup phase: SA first selects two large primes (e.g., more than 512 bits) P and Q, and computes the composite of P and Q, i.e. $N = P \cdot Q$. Then, SA determines a generator g modulo N with the order R, where R is a prime and is large enough (e.g., more than 160 bits) to withstand the exhaustive search attack. Finally, SA determines a one-way function f, where $0 < f(x) < R$ for any x. The one-way function f can be easily constructed from the current available hash functions such as MD5 or SHA-1 . At the end of this phase, SA publishes N and f, while keeping P, Q and R secret.

User registration phase: Suppose that the user U_i, with the identity ID_i and the password PW_i, wants to register with the system. It is assumed that U_i and SA have already applied the password authentication scheme to share the value of PW_i in advance. Denote $f(x)^{-1}$ as the inverse of $f(x)$ modulo R. Upon receiving the registration request originated from U_i, SA first randomly chooses an integer $d_i \in Z_R$ and then computes:

$$c_i = d_i{}^{f(ID_i,PW_i)^{-1}} modN$$
$$w_i = (g^{f(ID_i,PW_i)\cdot d_i} - f(ID_i))^{f(ID_i)^{-1}} modN \quad (1)$$

After that, SA sends $\{c_i,\ w_i\}$ to U_i via the public channel. Note that $\{c_i,\ w_i\}$ is self-certified, so that U_i can use his own password PW_i to compute $d_i = c_i{}^{f(ID_i,PW_i)} modN$ and verify the authenticity of $\{c_i,\ w_i\}$ by testing the equality:

$$g^{f(ID_i,PW_i)\cdot d_i} \stackrel{?}{=} (w_i^{f(ID_i)} + f(ID_i)) modN$$

At the end of this phase, U_i keeps $\{c_i,\ w_i\}$ in the memory of the computing device (e.g., IC cards) he holds.

Key agreement phase: Suppose that two registered parties U_i and U_j want to exchange an authenticated session key . They perform the following steps:
Step 1. U_i randomly chooses x_i and t_i bounded by the output length of f and computes:

$$d_i = c_i\ ^{f(ID_i, PW_i)} mod N$$

$$y_i = g^{x_i} mod N \quad (2)$$

$$r_i = g^{t_i} mod N \quad (3)$$

$$s_i = y_i \cdot t_i + r_i \cdot x_i + f(ID_i, PW_i) \cdot d_i \quad (4)$$

Then, U_i sends $\{ID_i, w_i, y_i, r_i, s_i\}$ to U_j.
Step 2. U_j verifies the authenticity of $\{ID_i, w_i, y_i, r_i, s_i\}$ by testing the equality:

$$g^{s_i} \stackrel{?}{=} r_i^{y_i} \cdot y_i^{r_i} \cdot (w_i^{f(ID_i)} + f(ID_i))(mod N)$$

Then, U_j computes y_j, r_j and s_j as that in Step 1, and computes the session key :

$$K_{j,i} = y_i^{x_j} mod N \quad (5)$$

and $m_j = f(ID_j, K_{j,i}, y_i)$. After that, U_j sends $\{ID_j, w_j, y_j, r_j, s_j, m_j\}$ to U_i.
Step 3. U_i verifies the authenticity of $\{ID_j, w_j, y_j, r_j, s_j\}$ as that in step 2. Then, U_i computes the session key $K_{i,j} = y_j^{x_i} mod N$ and tests the equality: $f(ID_j, K_{i,j}, y_i) \stackrel{?}{=} m_j$. If the equality holds, U_i is confirmed that he shares the same session key with U_j. After that, U_i computes $m_i = f(ID_i, K_{i,j}, y_j)$ and sends $\{ID_i, m_i\}$ to U_j.
Step 4. U_j tests the equality: $f(ID_i, K_{j,i}, y_j) \stackrel{?}{=} m_i$. If the equality holds, U_j is confirmed that he shares the same session key with U_i.

3. Proposed n-PAKA Protocol

The proposed n-PAKA protocol is like the CLIQUES multiparty key agreement protocol proposed by Steiner, et al. [Steiner *et al.*, 1997]. Suppose that n registered parties U_i ($1 \leq i \leq n$) want to achieve mutual authentication and key agreement. First, they determine the communication sequence, $U_1, U_2, ..., U_n$. Let U_1 act as the originator of the n-PAKA protocol. They perform the following steps:

Step 1: U_1 randomly chooses x_1 and t_1 bounded by the output length of f, obtains y_1 and r_1 as that in step 1 of 2-PAKA and computes

$$s_1 = y_1 \cdot t_1 + r_1 \cdot x_1 + f(X_1) \cdot f(ID_1, PW_1) \cdot d_1 \quad (6)$$

where $X_1 = \{g, g^{x_1}\}$. After that, U_1 sends $\{ID_1, X_1, w_1, y_1, r_1, s_1\}$ to U_2.
Step 2: For each $U_i(2 \leq i \leq n-1)$ receiving $\{ID_{i-1}, X_{i-1}, w_{i-1}, y_{i-1}, r_{i-1}, s_{i-1}\}$ from U_{i-1}, he verifies the equality:

$$g^{s_{i-1}} \stackrel{?}{=} y_{i-1}{}^{r_{i-1}} \cdot r_{i-1}{}^{y_{i-1}} \cdot (w_{i-1}{}^{f(ID_{i-1})} + f(ID_{i-1}))^{f(X_{i-1})} (mod N)$$

If the equality holds, U_{i-1} is authenticated. Then, U_i obtains y_i, r_i, s_i, and X_i, as that in setp 1, where

$$X_i = \{(g^{\prod_{(k=1\&k\neq j)}^{i} x_k} modN | 1 \leq j \leq i), g^{\prod_{k=1}^{i} x_k} modN\}$$

After that, U_i sends $\{ID_i, X_i, w_i, y_i, r_i, s_i\}$ to U_{i+1}.
Step 3: U_n first verifies the authenticity of $\{ID_{n-1}, X_{n-1}, w_{n-1}, y_{n-1}, r_{n-1}, s_{n-1}\}$ as that in step 2 and obtains X_n, y_n, r_n and s_n as that in step 1, where $X_n = \{(g^{\prod_{(k=1\&k\neq j)}^{n} x_k} | 1 \leq j \leq n)\}$. Then U_n computes the group session key $K_n = (g^{\prod_{k=1}^{n-1} x_k})^{x_n} modN$ and m_n=$f(ID_1, ID_2, ..., ID_n, K_n, X_n)$. After that, U_n broadcasts $\{ID_n, X_n, w_n, r_n, s_n, m_n\}$ to the other parties $U_i(1 \leq i \leq n-1)$.
Step 4: Each $U_i(1 \leq i \leq n-1)$ verifies the authenticity of $\{ID_n, X_n, w_n, y_n, r_n, s_n, m_n\}$ as that in step 1. Then, U_i computes the group session key:

$$K_i = (g^{\prod_{k=1\&k\neq i}^{n} x_k})^{x_i} modN \qquad (7)$$

After that, U_i tests the equality: $m_n \stackrel{?}{=} f(ID_i, ID_2, ..., ID_n, K_i, X_n)$. If the equality holds, it means that $K_i = K_n$ and U_i is confirmed that he shares the same group session key with U_n.

4. Security Analysis

The security of the proposed PAKA protocols is based on the intractability of computing discrete logarithm modulo a large composite (DLMC) [McCurley, 1988]. That is, given a large composite, N, of two primes, P and Q, a generator g over Z_N, and $y = g^x modN$, it is computationally infeasible to find x.

Achievement of the known-key security. From equations 5 and 7, it is to see that each U_i will produce a unique session key that depends on different x_i. By equations 2, 3, 4 and 6, an attacker will face the DLMC problem to compute x_i from public information $\{y_i, r_i, s_i\}$.

Achievement of the perfect forward secrecy. Suppose that PW_i is compromised. From equations 2, 3, 4 and 6, it can be seen that an attacker will face DLMC problem to compute x_i from PW_i and $\{y_i, r_i, s_i\}$.

Resistant of on-line password guessing attacks. An attacker cannot ensure the success of launching the on-line guessing attack, since no trusted server is required during the key agreement phase.

Resistant to off-line password guessing attacks. ¿From equations 1, 4 and 6, it can be seen that an attacker will face the DLMC problem for working out $f(ID_i, PW_i)$ and then off-line guessing PW_i from $\{w_i, y_i, r_i, s_i\}$.

Resistant to password-compromised impersonation attacks. For the success of launching such attacks, an attacker should have the ability to construct $\{ID_i, w_i', y_i', r_i', s_i'\}$ satisfying

$$w_i' = ((g^{s_i'}/y_i'^{r_i'} \cdot r_i'^{y_i'} - f(ID_i))^{f(ID_i)^{-1}} modN$$

However, the attacker will face the DLMC problem for working out $f(ID)^{-1}modR$.

Resistant to unknown key-share attacks. For the success of launching such attacks, an attacker should have the ability to construct $\{ID_a, w_a, y_i, r_i, s_a\}$ satisfying

$$s_a = y_i \cdot t_i + x_i \cdot r_i + f(ID_a, PW_a) \cdot d_a$$

However, the attacker will face the DLMC problem for working out x_i from $g^{x_i} modN$

5. Conclusion

We have presented a 2-PAKA protocol based on self-certified approach. The proposed 2-PAKA protocol can be further generalized to an n-PAKA protocol. Under the DLMC assumption, the proposed PAKA protocols achieve the security requirements of perfect forward secrecy and known-key security. Besides, the proposed PAKA protocols are resistant to the well-known attacks, such as on-/off-line password guessing, password-compromised impersonation, and unknown key-share.

References

[Bellare *et al.*, 2000] M. Bellare, D. Pointcheval and P. Rogaway (2000) Authenticated Key Exchange Secure Against Dictionary Attacks. *Advances in Cryptology – EUROCRYPT 2000, Lecture Notes in Computer Science 1807*, pages 139-155.

[Bellovin and Merritt, 1992] S. Bellovin and M. Merritt (1992). Encrypted Key Exchange: Password-based Protocols Secure Against Dictionary Attack. *Proceedings IEEE Symposium on Research in Security and Privacy*, pages 72-84.

[Blake-Wilson and Menezes, 1998] S. Blake-Wilson and A. Menezes (1998). Authenticated Diffie-Hellman Key Agreement Protocols . *Proceedings of the 5th Annual Workshop on Selected Areas in Cryptography – SAC'98*, pages 339-361.

[Boyko *et al*, 2000] V. Boyko, P. Mackenzie and S. Patae (2000). Provably-secure Password Authentication and Key Exchange Using Diffie-Hellman. *Advances in Cryptology – EUROCRYPT 2000, Lecture Notes in Computer Science 1807*, pages 156-171.

[Bresson *et al.*, 2000] E. Bresson, O. Chevassut, D. Pointcheval, and J.J. Quisquater (2001). Provably Authenticated Group Diffie-Hellman Key Exchange. *ACM Conference on Computer and Communications Security*, pages 255-264.

[Ding and Horster, 1995] Y. Ding and P. Horster (1995). Undetectable On-line Password Guessing Attacks. *ACM Operating System Review*, Vol. 29, No. 4, pages 77-86.

[Lee *et al.*, 1999] H. Lee, K. Sohn, H. Yang, and D. Won (1999). The Efficient 3-pass Password-based Key Exchange Protocol with Low Computational Cost for Client. *The Second International Conference Information Security and Cryptology – ICISC'99*, pages 147-155.

[Lin *et al.*, 2001] C.L. Lin, H.M. Sun, M. Steiner and T. Hwang (2001). Three-party Encrypted Key Exchange without Server Public-Keys. *IEEE Communication Letters*, Vol. 5, No. 12, pages 497-499

[McCurley, 1988] K. McCurley (1988). A Key Distribution System Equivalent to Factoring. *Journal of Cryptology*, Vol. 1, No. 19, pages 95-105.

[MacKenzie *et al.*, 2000] P. MacKenzie, S. Patel and R. Swaminathan (2000). Password-Authenticated Key Exchange Based on RSA . *Advances in Cryptology – ASIACRYPT 2000, Lecture Notes in Computer Science 1876*, pages 599-613.

[Steiner *et al.*, 1995] M. Steiner, G. Tsudik and M. Waidner (1995). Refinement and Extension of Encrypted Key Exchange. *Operating System Review*, Vol. 29, No. 3, pages 22-30.

[Steiner *et al.*, 1997] M. Steiner, G. Tsudik and M. Waidner (1997). CLIQUES: A New Approach to Group Key Agreement . *Technical Report RZ 2984, IBM Research.*

II

P2P AND AD HOC NETWORKS

Chapter 6

EXPERIMENTING WITH ADMISSION CONTROL IN P2P NETWORKS

Nitesh Saxena
Computer Science Department
University of California at Irvine
nitesh@ics.uci.edu

Gene Tsudik
Computer Science Department
University of California at Irvine
gts@ics.uci.edu

Jeong Hyun Yi*
Computer Science Department
University of California at Irvine
jhyi@ics.uci.edu

Abstract Peer-to-peer (P2P) security has received a lot of attention as of late. Most prior work focused almost entirely on issues related to secure communication, such as key management and peer authentication . However, an important pre-requisite for secure communication – secure peer admission – has been neither recognized nor adequately addressed. Only very recently, some initial work began to make inroads into this difficult problem. In particular, [Kim et al., 2003] constructed a peer group admission control framework based on various admission policies matched with appropriate cryptographic techniques. Recent results [Saxena et al., 2003, Narasimha et al., 2003] also illustrate the design of, and experiments with, certain group admission control mechanisms.

In this work, we report on the implementation of Bouncer, an experimental peer group admission control toolkit used in [Saxena et al., 2003] and its trial integration with two peer group systems with very different goals and seman-

*correspondence author

tics: Gnutella and Secure Spread . We also discuss some outstanding issues, challenges and future research directions relevant to this topic.

Keywords: Access Control , Peer-to-Peer Networks, Peer Group Communications

1. Introduction

The rising popularity of P2P applications prompts the need for specialized P2P security services and mechanisms. This has been recognized by the research community, however, the bulk of prior work is concerned with secure P2P communication, e.g., authentication, anonymity and key management . Although these issues are certainly important, another equally important topic has remained mostly unaddressed. Informally, it has to do with how one becomes a peer in a P2P system. More concretely, the technology for secure admission of peers into a P2P application simply does not exist. This statement does not contradict the fact that there are many currently operating P2P applications; they either operate in a completely open manner (i.e., have no admission control whatsoever) or admit peers on some *ad hoc* basis. This state of affairs bears a certain similarity to the early days of group key management when group keying was either non-existent or obtained by out-of-band means. To exploit this a little further, we observe that, just as trivial key management solutions severely limited the functionality of peer group applications, equally trivial admission control techniques will do (or already have done) the same. In other words, we believe that – without a well-thought-out architecture and appropriate techniques for peer admission – most P2P systems will sooner or later hit the proverbial "brick wall".

1.1 Prior Work

Recently, [Kim et al., 2003] developed a group admission control framework based on various cryptographic techniques. This framework classifies group admission policy according to the entity (or entities) that makes peer admission decisions. The classification includes simple admission control policies, such as static ACL(Access Control List)- or attribute-based admission, as well as admission based on the decision of some fixed entity: external (e.g., a TTP) or internal (e.g., a group founder). Such simple policies are relatively easy to support and do not present much of a technical challenge. However, they are inflexible and ultimately unsuitable for a dynamic P2P setting. Static ACLs enumerate all possible members and hence cannot support truly dynamic membership (although they work well for closed groups). Admission based on decisions of a TTP or a group founder violates the peer nature of P2P, since the entire philosophy of P2P paradigm is based on collective, distributed services and decisions.

To address more challenging collective (group-centric) admission policies, a follow-on work [Saxena et al., 2003] built upon the framework in [Kim et al., 2003] by designing a menu of suitable distributed mechanisms on a number of cryptographic techniques. This work yielded mechanisms for both centralized and (more challenging, yet also more realistic) decentralized group settings. In the latter, all current group members can take part in the admission process in a fully distributed manner. This work also assessed the practicality of distributed cryptographic mechanisms (such as verifiable threshold signatures) in both synchronous and asynchronous P2P settings. For an in-depth discussion of these admission control mechanisms, protocols and the experimental results, the reader is referred to [Saxena et al., 2003, Narasimha et al., 2003, Kim et al., 2003].

In this work we focus on the design and implementation of Bouncer, the admission control toolkit [Saxena et al., 2003] integrated with an asynchronous P2P system (Gnutella [Gnutella]) and a synchronous group communication system with strong membership semantics (Secure Spread [SSPR]). The Bouncer toolkit is general, i.e., it can be easily grafted onto any peer group setting.

2. Background

In this section, we describe a typical P2P admission procedure. The goal of this procedure is to allow a prospective member to obtain a group membership certificate. Using this certificate, a new member can prove membership and take part in future admission decisions.

As described in [Saxena et al., 2003], the admission process is similar to a general voting mechanism whereby a prospective member needs to collect a certain minimum (threshold) number of positive votes (endorsements) before becoming a group member. There are two types of threshold admission policies: fixed and dynamic. The former is specified as the minimum number of votes, whereas, a dynamic threshold is specified as a fraction or percentage of the current group size. A fixed threshold is essentially a t-out-of-n model where the threshold t is fixed and n (current group size) varies over time. In contrast, a dynamic threshold (such as 30%) implies that t shrinks or grows in tandem with n.

The table below summarizes the notation used in the remainder of the paper. The "generic" peer admission process is as follows:

Step 0. Bootstrapping: A prospective peer M_{new} obtains the *group charter* [Kim et al., 2003] out of band and then the information of current group size from either GAuth or some bootstrap node. The group charter contains

TD	trusted dealer
GAuth	group authority
n	total number of peers
t	threshold ($t \leq n$)
M_{new}	prospective member
M_i	current member ($0 < i \leq n$)
PKC_{new}	public key certificate of M_{new}
GMC_{new}	group membership certificate of M_{new}

various parameters and admission policies, including: group name, signature/encryption algorithm identifiers, threshold (numeric or fractional corresponding to fixed or dynamic threshold, respectively), below-threshold policy and other optional fields. This process is performed only once per admission.

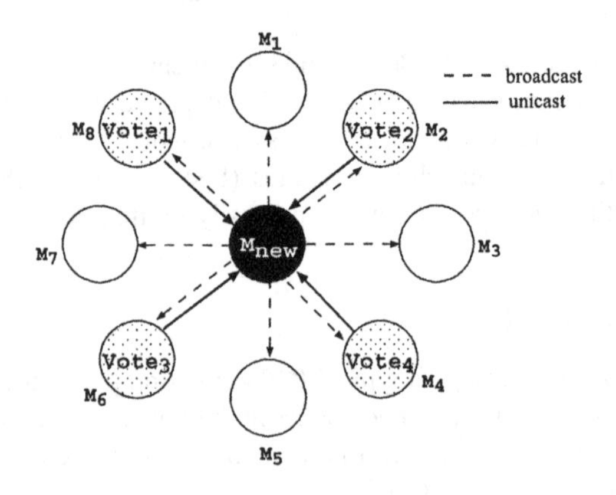

Figure 6.1. Admission Control

Step 1. Join Request: As shown in Figure 6.1, M_{new} initiates the protocol by sending a join request (JOIN_REQ) message to the group. This message, signed by M_{new}, includes M_{new}'s public key certificate (PKC_{new}) and the target group name. How this request is sent to the group is application-dependent.[1]

Step 2. Admission Decision: Upon receipt of JOIN_REQ, a group member first extracts the sender's PKC_{new} and verifies the signature. If a voting peer approves of admission it replies with a signed message (JOIN_COMMIT) Several signature schemes (as described later in this section) can be used for this purpose. M_{new} verifies each vote.

Step 3. GMC Issuance: Exactly who issues the GMC_{new} for M_{new} depends on the security policy. If the policy stipulates using an existing GAuth, once enough votes are collected (according to the group charter), M_{new} sends to the

GAuth a group certificate request message (GMC_REQ). It contains: PKC_{new}, group name, and the set of collected votes. In a distributed setting with no GAuth, M_{new} verifies the individual votes, and, from them, composes her own GMC_{new}.

Armed with a GMC , M_{new} can act as a *bona fide* group member. To prove membership to another party (within or outside the group) M_{new} simply signs a message (challenge) to that effect.

To perform the admission decision process, various signature schemes are used, namely the plain RSA , Threshold RSA (TS-RSA) [Kong et al., 2001, Luo et al., 2002, Kong et al., 2002], Threshold DSA (TS-DSA) [Gennaro et al, 1996] and Accountable Subgroup Multisignatures (ASM) [Ohta et al., 2001]. For a detailed description of these signature schemes and the admission protocol, refer to [Saxena et al., 2003] and [Narasimha et al., 2003].

3. Bouncer: Admission Control Toolkit

We have implemented Bouncer, a general-purpose toolkit for P2P admission control based on the description in Section 2. All cryptographic functions are developed using the OpenSSL library [OpenSSL]. The toolkit is written in C on Linux and currently consists of about $45,000$ lines of code. The source code for the membership control toolkit is publicly available at [PGAC].

3.1 System Design

The admission control system is made up of three basic layers of the architecture; GAC APIs, security and management services, and the underlying cryptographic functions. Figure 6.2 illustrates the architecture.

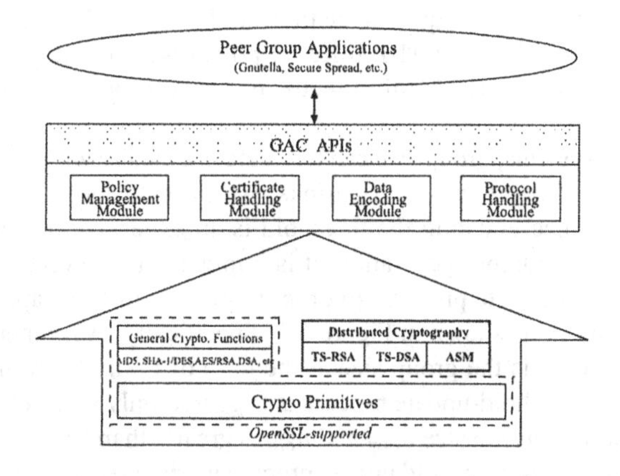

Figure 6.2. GAC System Architecture

The GAC APIs define the application programming interface for accessing the admission control services. These APIs are useful when integrating our Bouncer with other peer group applications. The security and management services are carried out by the following modules:

- Policy Management Module
- Certificate Handling Module
- Data Encoding Module
- Protocol Handling Module

All security services are provided by the underlying cryptographic libraries.

3.2 Cryptographic Libraries

Most of the general cryptographic functions such as SHA-1 , RSA , DSA, and so on, are supported by OpenSSL . Specifically, we have implemented three distributed cryptographic schemes on top of OpenSSL, and embedded our libraries into it. The Bouncer supports four different signature schemes; plain RSA , ASM , TS-RSA, and TS-DSA as addressed earlier.

3.3 Security and Management Services

3.3.1 Policy Management Module. A *policy management* module is the component which checks for conformance to the policy specified in the *group charter* [Kim et al., 2003]. First, this module contains functions to check the threshold type. If the threshold type is a static, it checks if the number of current members is at least equal to the threshold ($n \geq t$). If $n < t$, the policy manager enforces the `BelowThreshold` policy which requires it to either forward the `JOIN_REQ` to GAuth directly, or to reset the threshold to reflect current n.

In most P2P systems, group size can fluctuate drastically within a short time. As the number of peers grows or shrinks, we need to increase or decrease the threshold. Since updating the threshold is an expensive operation which requires a random number generation, it is impractical for every membership event to trigger an update process. In order to prevent this, we apply a simple *window* mechanism as shown in Figure 6.3. Specifically, every member keeps state of n_{old}, which is the group size at the time of the last threshold-update process. A new threshold-update process is triggered only when the difference between the current group size n_{cur} and n_{old} is greater than Win – the window buffer. In other words, threshold update process is triggered only when $|n_{cur} - n_{old}| \geq Win$.

```
Function GAC_Dynamic_Threshold_Update ();

Input parameters:
    X509* GChart,          \* group charter *\
    int N_old,             \* old group size *\
    int N_cur,             \* current group size *\
    int T_cur              \* current threshold *\

Body:
    int diff;
    int offset;
    int Win;               \* Window buffer size *\
    int T_new              \* new threshold *\

    T_new=T_cur;
    Win=WIN_TIMES*GChart.threshold.fixed;
    diff = N_cur - N_old;
    if (diff >= Win) {
        offset = ⌊diff / Win⌋;
        N_old = N_old + (offset*Win);
        T_new = ⌊(GChart.threshold.dynamic /
                  100) * N_old⌋;
        if (T_new > T_cur)
            T_cur=T_new;
    }
    return T_new;
```

Figure 6.3. Dynamic Threshold Update Procedure

3.3.2 Certificate Handling Module. Both GMC -s and group charters generated by the Bouncer are compatible with X.509v3 [Housley et al., 2002]. This *certificate handling* module takes care of all functions related to certificate compatibility. For example, in the group charter, we need to define several attributes in the extension field of certificate in order to codify certain admission policy. And this module also has a function to bind the identity of GMC to that of PKC as shown in Figure 6.4 to protect against the Sybil attack [Douceur, 2002], assuming that a Certification Authority (CA) issues a PKC with a unique identity to each user.

Further, possession of a GMC does not prove that the GMC actually belongs to the bearer. One way to accomplish this is by requiring for every group member to have a standard X.509 public key certificate (PKC) issued by the CA. The GMC simply needs to contain the public key of the member extracted from her PKC. Now the member (bearer of a GMC) can prove ownership of the GMC by demonstrating knowledge (e.g., by signing a message) of the private key corresponding to the public key referred to in in the GMC.

3.3.3 Data Encoding Module. The *data encoding* module contains all encoding and decoding functions which convert ASN.1-formed messages to and from DER-encoded from. For example, `i2d_PS_Join_Request()` is a function which converts ASN.1-structured `JOIN_REQ` message based on plain RSA into DER-encoded binary data in order to transfer the message over the

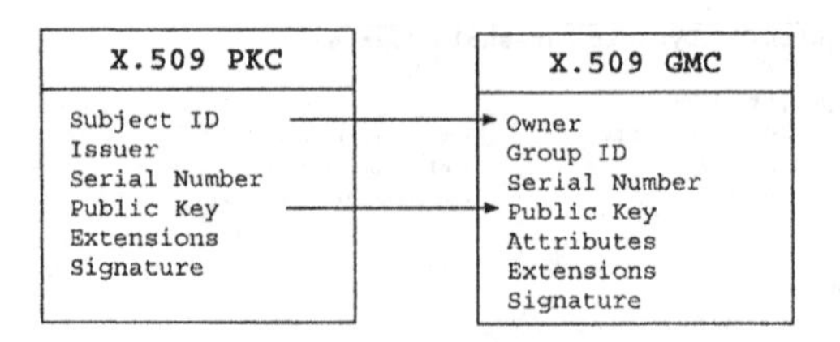

Figure 6.4. Binding GMC to PKC

networks. Similarly, `d2i_PS_Join_Request()` is called when receiving `JOIN_REQ` message, to get internal form of the message.

3.3.4 Protocol Handling Module. The *protocol handling* modules includes functions used to identify admission control protocols and transfer the messages to and from the corresponding libraries. Figure 6.5 shows the structure of GAC packet. Each packet is classified on the protocol using the packet type in the packet header.

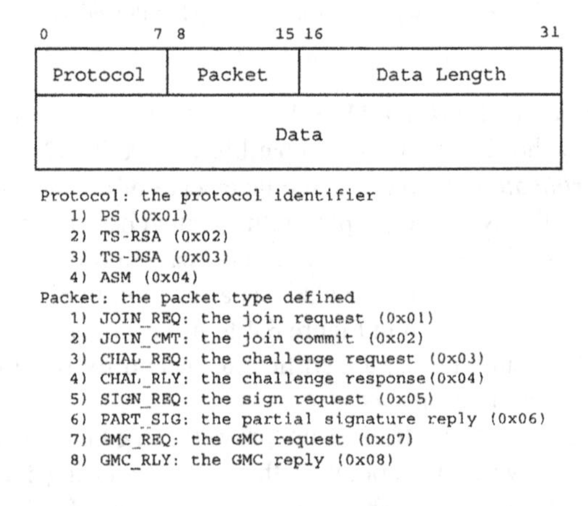

Figure 6.5. GAC Packet Structure

3.4 GAC APIs

Application developers require no special knowledge of the organization of the security and management modules as well as cryptographic libraries. They just need to use the GAC function interface to build any application. GAC APIs are logically partitioned into functional categories. The goal of this logical partitioning is to assist application developers in understanding and making effective use of the security APIs. With this logical classification, we support the following APIs. Among these APIs, `GMC_Request()` and `GMC_Reply()` are optionally required only when we can assume the presence of a centralized authority.

- Plain RSA APIs

```
GAC_PACKET *PS_Join_Reqest();
GAC_PACKET *PS_Join_Commit();
GAC_PACKET *PS_GMC_Request(); /* optional */
GAC_PACKET *PS_GMC_Reply();   /* optional */
```

- TS-RSA APIs

```
GAC_PACKET *TSS_Join_Request();
GAC_PACKET *TSS_Join_Commit();
GAC_PACKET *TSS_Sign_Request();
GAC_PACKET *TSS_Part_Sign();
GAC_PACKET *TSS_GMC_Request(); /* optional */
GAC_PACKET *TSS_GMC_Reply();   /* optional */
```

- TS-DSA APIs

```
GAC_PACKET *TSD_Join_Request();
GAC_PACKET *TSD_Join_Commit();
GAC_PACKET *TSD_Chal_Req();
GAC_PACKET *TSD_Chal_Rly();
GAC_PACKET *TSD_Rnd_Req();
GAC_PACKET *TSD_Rnd_Rly();
GAC_PACKET *TSD_Sign_Request();
GAC_PACKET *TSD_Part_Sign();
GAC_PACKET  *TSD_GMC_Request(); /* optional */
GAC_PACKET *TSD_GMC_Reply();    /* optional */
```

- ASM APIs

```
GAC_PACKET *ASM_Join_Request();
GAC_PACKET *ASM_Join_Commit();
GAC_PACKET *ASM_Sign_Request();
GAC_PACKET *ASM_Part_Sign();
GAC_PACKET *ASM_GMC_Request(); /* optional */
GAC_PACKET *ASM_GMC_Reply();   /* optional */
```

4. Integration with Peer Group Systems

To evaluate the performance of our mechanisms and to measure the overhead incurred due to incorporating admission control in the context of real-world application, we integrated the Bouncer with a popular P2P file sharing system, Gnutella /indexGnutella and with a wide area secure group communication system, Secure Spread . Secure Spread is selected as an example of a synchronous P2P system, and Gnutella as an asynchronous one. We integrated the centralized admission protocol with the former and the decentralized one with the latter to measure the performance in both settings.

In the following sub-sections, we discuss the implementation details for the integration with both the systems.

4.1 Integration with Gnutella

The *Gnutella* is the "pure" P2P file sharing system which is closest to the ideal structure of the P2P spirit, where all participants have uniform role. In such an architecture, users are free to join and leave the group. Even malicious users can easily join to deny or disrupt the system. To prevent such a security threat in a fully distributed P2P environment, we integrated our Bouncer with Gnut-0.4.21 [Gnut] (an open-source Gnutella [Gnutella] implementation).

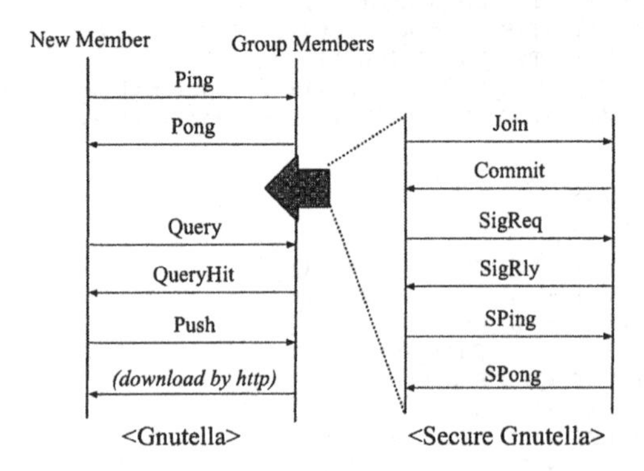

Figure 6.6. Secure Gnutella Protocol Flow

At the setup phase of the Gnutella protocol , a connection is established by communicating so-called `Ping` and `Pong` messages which are based on IP addresses as shown in Figure 6.6. To look for a file, a new member sends out a broadcast `Query` message to every member to which it is directly linked. The group members identifying the requested file in their repository answer with a `QueryHit` message which is returned to the connection from which the request arrived. The `QueryHit` message contains the *ResultSet* and the pair ⟨*IP address, port*⟩ that must be used to download the file via HTTP.

The *Secure Gnutella* protocol, illustrated in Figure 6.6, defines some extra messages for secure admission control ; `Join`, `Commit`, `SigReq`, `SigRly`, `SPing`, and `SPong`. The message format for new protocol steps is defined as follows;

- `Join(mesg, PKC, Sig)`
- `Commit(port, IP addr, GMC, commit_val, Sig)`

- `SigReq(servant ID, sigreq_val, Sig)`
- `SigRly(servant ID, sigrly_val, Sig)`
- `SPing(Group ID, GMC)`
- `SPong(port, IP adddr, # of files, # of Kbytes, GMC)`

First, like in a standard Gnutella protocol , a new member broadcasts to all her neighbors `Join` message which contains the join request message and her own PKC . Upon reception of the `Join` message, some of group members reply with `Commit` message to confirm that they will participate in admission process. In this message, the `commit_val` is an encapsulated message of the GAC protocol , which is DER-encoded form . The `SigReq` and `SigRly` are newly specified messages for the GAC protocol . For checking the integrity of protocol message, `Commit`, `SigReq`, and `SigRly` messages include the signature thereon which is PKCS7-formatted .

In order to prevent Sybil attacks [Douceur, 2002], we modified standard `Ping` and `Pong` messages so that the connection is made only if the responder answers with its valid GMC . For this purpose, we specified two new messages: `SPing` and `SPong`. The `SPing` message contains the requester's PKC , and the `SPong` message contains the responder's GMC and its signature (to prove possession of its private key). In *Secure Gnutella* system, standard `Ping` and `Pong` messages are no longer used.

4.2 Integration with Secure Spread

Spread [Spread] is a wide area group communication system . It provides a high performance messaging service that is resilient to faults across external or internal networks. Spread functions as a unified message bus for distributed applications, and provides highly tuned application-level multicast and group communication support. Spread services range from reliable message passing to fully ordered messages with delivery guarantees, even in case of computer failures and network partitions.

Secure Spread [SSPR] is an application built atop Spread. It enhances Spread by integrating security services and key management .

In its present form, Secure Spread supports only static group access control which is provided at the daemon level using ACL's. This clearly poses a single point of failure problem. Moreover, as argued before, static admission control is no good for dynamic groups. Secure Spread also has a notion of a *flush* mechanism, in which all current group members need to acknowledge any change in membership (e.g. join, leave, partition, merge). A prospective member can join a group only after it has received *flush OK* messages from all current group members. This is a very weak form of providing admission as this mechanism offers no security at all because there involves no authentication of either prospective or current members. Moreover, all group members need to be involved in every admission process simultaneously.

In order to resolve these problems and of course to measure the performance, we integrated Bouncer with Secure Spread . The integration involves extension to the Spread API and can be used with any application (including Secure Spread) that uses Spread.

We added the following function to the current interface of Spread.

```
int SP_GAC_join(mailbox mbox, const char *group)
```

This function is declared in `sp.h` of Spread source tree. It joins a group using the group admission mechanisms described in previous sections, with the name passed as the string group. If the group does not exist among the Spread daemons it is created, otherwise it joins the existing group. The mbox of the connection upon which to join a group is the first parameter. The group string represents the name of the group to join.

The function Returns 0 on success or one of the following errors (< 0):

`ILLEGAL_GROUP`
The group given to join was illegal for some reason. Usually because it was of length 0 or length > `MAX_GROUP_NAME`.

`ILLEGAL_SESSION`
The session specified by mbox is illegal. Usually because it is not active.

`CONNECTION_CLOSED`
During communication errors occurred and the join could not be initiated.

In case, the prospective member is not able to receive enough votes, the function call will not be completed and the member will wait forever.

`JOIN_REQ` message is encapsulated within the standard spread message and sent to all the group members using Spread multicasting. Figures 6.7 show Spread header and the encapsulation of GAC message inside the spread message (sizes are in bytes). The function makes a call to the `SP_multicast` function of the Spread API . For details regarding the multicast message, refer to the spread function interface in [Spread].

In order to receive replies back from the group members, the function `SP_GAC_join()` uses the `SP_receive` function of the Spread API .

We have also modified the `SP_receive` function. This takes care of the fact that when a current group member receives the `JOIN_REQ` message from a prospective member, it responds with a `JOIN_CMT` message as its vote. This message again is encapsulated within the standard Spread message and its sent to the requesting member using the Spread unicasting. For this purpose,

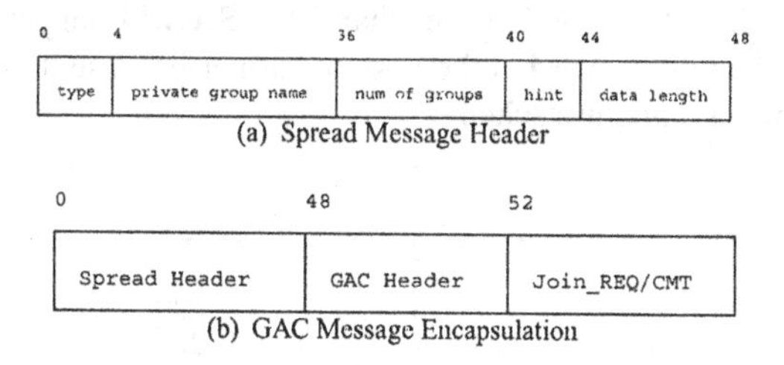

Figure 6.7. Spread GAC Message Encapsulation

we again use the `SP_multicast` function to send unicast message to the new member using its private group name which is represented by *#private user name#daemon name.*

After collecting enough votes from group members, the prospective member requests the GMC from the external GAuth. Once, the GAuth issues the GMC to the new member, the admission process is completed. Then, the spread daemons update the membership information and update/distribute the new key to the newly joined member.

5. Experiments

In our experiments with Gnutella and Secure Spread, we measured the costs of basic operations and then compared the performance of four cryptographic protocols with both fixed and dynamic thresholds. We used 1024-bit modulus in all mechanisms; that is, 1024-bit N in RSA and TS-RSA, and 1024-bit p and 160-bit q in TS-DSA and ASM.

Since each protocol has different number of communication rounds, we measured total processing time from sending of the `JOIN_REQ` to obtaining new GMCs[2]. This means the join cost includes not only the signature generation and verification time in basic operations, but also the communication costs such as packet encoding/decoding time, the network delay, and so on. To get reasonably correct results, the experiments were repeated more than 1000 times for each.

5.1 Computation Costs

In this section, we demonstrate the cost of each signature scheme used as a primitive in Bouncer.

Figure 6.8(a) shows the cost of signature generation versus the key size, where t=3. We found that in TS-RSA , the cost in generating a signature is much more expensive than that of RSA signature generation, since we can not apply *CRT (Chinese Remainder Theorem)* to speed up the computation as in plain RSA scheme. TS-RSA is slightly better than TS-DSA with 512-

bit modulus, while TS-DSA is faster than TS-RSA with larger key size. As evident from the figure, ASM is the best performer because it is based on the efficient Schnorr's signature scheme.

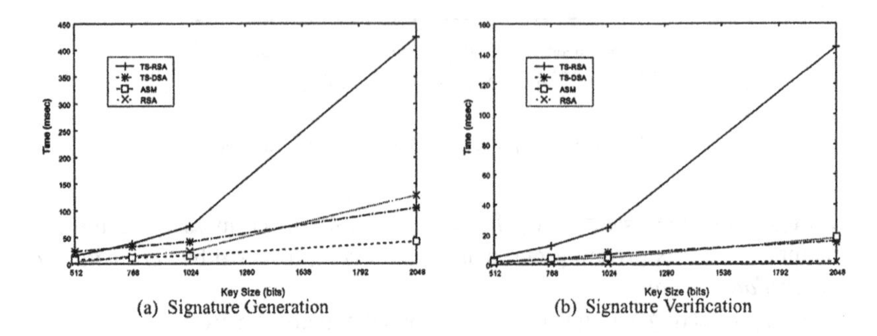

(a) Signature Generation (b) Signature Verification

Figure 6.8. Basic Operation Cost

Figure 6.8(b) shows the cost of signature verification with varying key sizes. In PS, the cost of signature verification is proportional to the threshold. All other schemes, except PS, have only one resulting signature due to the aggregation of partial signatures. We also observe that the verification costs of TS-DSA and ASM are almost the same as for the underlying DSA and Schnorr signature schemes respectively. However, verification cost for TS-RSA is extremely high. This is because $m^N \pmod N$ in *t-bounded offsetting algorithm* [Kong et al., 2001] has to be computed almost every time the signature is verified. Due to this expensive operation, it turns out that the TS-RSA performs much worse than the other schemes, contrary to our intuition.

5.2 Signature Size

From the analysis of the computation cost above, it turned out that both plain RSA and ASM are more efficient than the two threshold signature schemes. However, the length of the signature in plain RSA and ASM is linear in threshold t. In this experiment we extract the identities (which are X.509 DN formatted) from the GMC -s. We also used 1024-bit RSA key and SHA-1 as a hash function for both ASM and TS-DSA .

In both plain RSA and ASM schemes, the signers' identities should be included in the resulting signature. Due to the size of the identity (i.e., 952 bits), the resulting signatures become very large depending on the threshold; whereas, both TS-RSA and TS-DSA have a constant signature size (i.e., 1024 bits and 320 bits, respectively). For example, from the Figure 6.9, we can see that the size of plain RSA is about 150 times as long as that of ASM when the

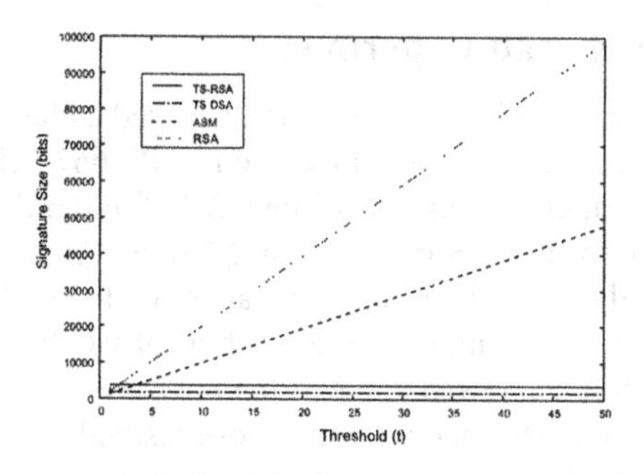

Figure 6.9. Signature Size

threshold is set to 25. Therefore, we recognize that both plain RSA and ASM would not be suitable for large groups where the bandwidth is a major concern.

5.3 Gnutella Experiments

We measured the performance of the `Secure Gnut` which is the *Gnut* system integrated with our Bouncer. We performed all measurements on the following Linux machines connected with a high-speed LAN: P4-1.2GHz, P3-977MHz, P3-933MHz, and P3-797MHz.

Figure 6.10(a) shows the join cost for the static threshold case. Figure 6.10(b) shows the join costs for the dynamic threshold case where the threshold ratio is set to 30% of current group size. All of these measurements were performed with the equal number of member processes on each machine.

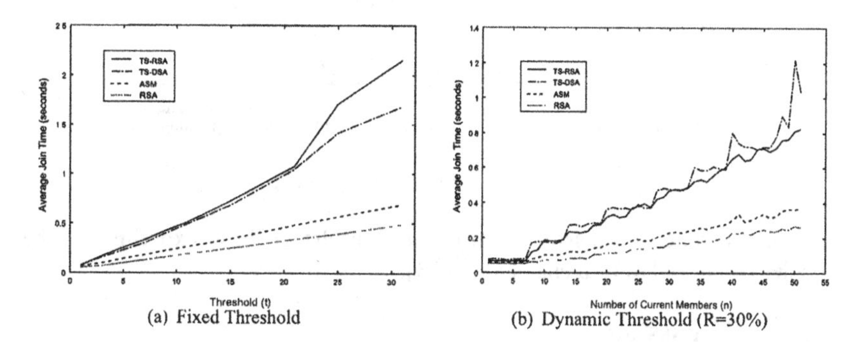

(a) Fixed Threshold

(b) Dynamic Threshold (R=30%)

Figure 6.10. Gnutella Experiments

5.4 Secure Spread Experiments

For our experiments with Secure Spread , we used a cluster of 10 machines at Johns Hopkins University. Each machine has P3-667 MHz CPU, 256 KB Cache and 256 MB memory and runs Linux 2.4. We ran Spread daemons on all machines which formed a Spread Machine Group. Almost equal number of clients running on these machines connect randomly to the daemons. The new joining member is a client running on a machine at UC Irvine with a Celeron 1.7 GHz CPU, 20 KB cache and 256 MB memory.

Experiments were performed with the above testbed for both fixed and dynamic thresholds for all signature schemes discussed thus far.

Figure 6.11(a) shows the plot for the average time taken by a new member to join a group with a fixed threshold. We performed this test with 4-5 processes on each machine and measured the join cost by changing the threshold. As expected, plain RSA is the best performer in terms of computation time. However, we also see that both TS-RSA and TS-DSA exhibit reasonable costs (< 1 sec.), at least until t=10.

Figure 6.11(b) show the plots for the average time for a new member to join a group with a dynamic threshold. In this experiment, the threshold ratio (R) is set to 30% of the current group size. The actual numeric threshold is determined by multiplying the group size by R. We measured the performance up to $n = 50$.

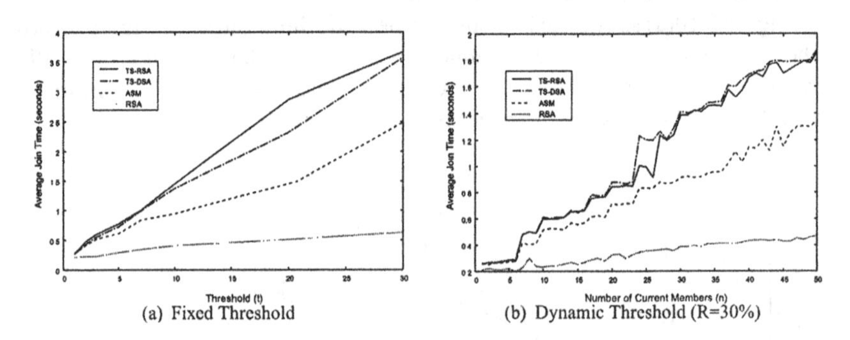

(a) Fixed Threshold (b) Dynamic Threshold (R=30%)

Figure 6.11. Secure Spread Experiments

For a detailed discussion regarding the results of these experiments, the reader is referred to [Saxena et al., 2003].

6. Discussion

As it is clearly reflected from the measurement results above, all the advanced cryptographic constructs i.e. the threshold signatures and the multisignatures perform quite poorly. Especially the threshold signatures are about 4-7 times costlier than the plain RSA signatures for relatively larger groups. But, as discussed in [Saxena et al., 2003], since for plain signatures and multisignatures the size of the combined signature and thus the size of the GMC varies proportionally with the threshold, we can't pick just one signature scheme for all P2P settings. A certain balance has to be maintained between the size of the GMC and the average join cost apart from the choice of the scheme-specific features like anonymity, accountability, membership awareness and so on.

One might argue that group signature scheme [Ateniese et al., 2000] might also be a possible candidate for the admission control especially in a P2P scenario where signer anonymity is a must. We did in fact implement the group signature scheme in our toolkit and experimented with it. But, unfortunately, we have to rule out the possibility of using group signatures as they perform way worse than the other signature schemes. Moreover, group signature scheme can only be used for the centralized admission protocol as it requires the presence of a group manager.

In summary, we are faced with a couple of challenges in order to provide secure admission control. One challenge is to make the admission process as distributed as possible and the other is to do so in a highly efficient manner with the lowest possible overhead (storage as well as bandwidth). Though in a P2P setting, a distributed approach seems like the most natural one but it turns out to be the hardest as well. A admission control mechanism will only be applicable in mobile ad-hoc and sensor networks if it is both distributed and power-efficient. As of now none of the schemes seem very useful in these scenarios.

7. Future Directions

As is evident from the experimental results and above discussion, there is a lot of scope for improvement and promise for further work. We have seen that there is a trade-off between the performance and the signature size among various schemes. So, one immediate objective is to find/design an efficient signature scheme which on one hand has a fewer rounds in the protocol and on the other smaller signature size in the GMC . Recently proposed *a*ggregated signature scheme [Boneh et al., 2003] appears to be an attractive candidate for the same. But, we claim that one particular signature scheme would not be sufficient for our purpose of admission control. The choice of the scheme to be used has to be made based on a number of factors like type/size of group, bandwidth, various features desired and the group policies.

Another possible enhancement could be to have admission decision based on a trust based model. In the usual more practical scenario, a group member can only probabilistically vote in or vote out a prospective member. In the presence of a trust model, voting would be more deterministic.

This work uses a certificate based approach toward admission control. With certificates arises the issue of revocation which could be a hard problem to deal with in a distributed setting. In order to avoid this issue, another future direction is to design a non-certificate based approach for admission.

Another prospect of future work is the complementary problem of membership revocation. If providing secure admission is hard, solving the problem of revocation will be even more challenging.

Notes

1. Note that PKC_{new} does not have to be an identity certificate; it could also be a group membership certificate for another group.
2. In these experiments we did not consider the *partial share shuffling* for both TS-RSA and TS-DSA.

References

[Ateniese et al., 2000] Ateniese, G., Camenisch, J., Joye, M., and Tsudik, G. (2000). A Practical and Provably Secure Coalition-Resistant Group Signature Scheme. In Bellare, Mihir, editor, *CRYPTO '00*, number 1880 in LNCS, pages 255–270.

[Boneh et al., 2003] Boneh, D., Gentry, C., Lynn, B., and Shacham, H. (2003). Aggregate and Verifiably Encrypted Signatures from Bilinear Maps. In Biham, Eli, editor, *EUROCRYPT '03*, number 2656 in LNCS, pages 416–432.

[Douceur, 2002] Douceur, J. R. (2002). The Sybil Attack. In *International Workshop on Peer-to-Peer Systems (IPTPS'02)*.

[Gennaro et al, 1996] Gennaro, R., Jarecki, S., Krawczyk, H. and Rabin, T. (1996). Robust Threshold DSS Signatures. In Maurer, Ueli, editor, *EUROCRYPT '96*, number 1070 in LNCS, pages 354–371.

[Gnutella] Gnutella Protocol Specification v0.4 (`http://www.clip2.com/GnutellaProtocol04.pdf`).

[Gnut] Gnut v0.4.21 source code (`http://schnarff.com/gnutelladev/source/gnut`).

[Housley et al., 2002] Housley, R., Polk, W., Ford, W., and Solo, D. (2002). Internet X.509 Public Key Infrastructure Certificate and Certificate Revocation List (CRL) Profile. RFC 3280, IETF.

[Kong et al., 2002] Kong, J., Luo, H., Xu, K., Gu, D. L., Gerla, M., and Lu, S. (2002). Adaptive Security for Multi-level Ad-hoc Networks. In *Journal of Wireless Communications and Mobile Computing (WCMC)*, volume 2, pages 533–547.

[Kim et al., 2003] Kim, Y., Mazzocchi, D., and Tsudik, G. (2003). Admission Control in Peer Groups. In *IEEE International Symposium on Network Computing and Applications (NCA)*.

[Luo et al., 2002] Luo, H., Zerfos, P., Kong, J., Lu, S., and Zhang, L. (2002). Self-securing Ad Hoc Wireless Networks. In *Seventh IEEE Symposium on Computers and Communications (ISCC '02)*.

[Kong et al., 2001] Kong, J., Zerfos, P., Luo, H., Lu, S., and Zhang, L. (2001). Providing Robust and Ubiquitous Security Support for MANET. In *IEEE 9th International Conference on Network Protocols (ICNP)*.

[Narasimha et al., 2003] Narasimha, M., Tsudik, G., and Yi, J. H. (2003). On the Utility of Distributed Cryptography in P2P and MANETs: The Case of Membership Control. In *IEEE International Conference on Network Protocol (ICNP)*, pages 336–345.

[Ohta et al., 2001] Ohta, K., Micali, S., and Reyzin, L. (2001). Accountable Subgroup Multisignatures. In *ACM Conference on Computer and Communications Security*, pages 245–254.

[OpenSSL] OpenSSL Project (`http://www.openssl.org/`).

[PGAC] Peer Group Admission Control Project (`http://sconce.ics.uci.edu/gac`).

[Spread] Spread Project (`http://www.spread.org/`).

[SSPR] Secure Spread Project (`http://www.cnds.jhu.edu/research/group/secure_spread/`).

[Saxena et al., 2003] Saxena, N., Tsudik, G., and Yi, J. H. (2003). Admission Control in Peer-to-Peer: Design and Performance Evaluation. In *ACM Workshop on Security of Ad Hoc and Sensor Networks (SASN)*, pages 104–114.

Chapter 7

ADAPTIVE RANDOM KEY DISTRIBUTION SCHEMES FOR WIRELESS SENSOR NETWORKS

Shih-I Huang
Department of Computer Science and Information Engineering
National Chiao Tung University

Shiuhpyng Shieh
Department of Computer Science and Information Engineering
National Chiao Tung University

S.Y. Wu
Department of Computer Science and Information Engineering
National Chiao Tung University

Abstract Wireless Sensor Networks (WSNs) are formed by a set of small devices, called nodes, with limited computing power, storage space, and wireless communication capabilities. Most of these sensor nodes are deployed within a specific area to collect data or monitor a physical phenomenon. Data collected by each sensor node needs to be delivered and integrated to derive the whole picture of sensing phenomenon. To deliver data without being compromised, WSN services rely on secure communication and efficient key distribution . In this paper, we proposed two key distribution schemes for WSNs, which require less memory than existing schemes for the storage of keys. The Adaptive Random Pre-distributed scheme (ARP) is able to authenticate group membership and minimize the storage requirement for the resource limited sensor nodes. The Uniquely Assigned One-way Hash Function scheme (UAO) extends ARP to mutually authenticate the identity of individual sensors. The two proposed schemes are effective for the storage of keys in a wireless sensor network with a large number of sensors.

1. Introduction

Wireless Sensor Network (WSN) [Akyilidiz et al., 2002, Estrin et al., 1999] is a kind of network composed of nodes associated with sensors. Each node has the characteristics of small size, limited power, low computation power and wireless access. The sensor node is responsible for collecting and delivering data over wireless network, and it is desirable to keep the delivered data confidential along the wireless transmission path from one node to another. [Tilak et al., 2002, Kong et al., 2001]

To ensure secure peer-to-peer wireless communication [Slijepcevic et al., 2002, He et al., 2003, Heinzelman et al., 1999, Intanagonwiwat et al., 2000, Zhou et al., 1999, Luo et al., 2002, Hubaux et al., 2001, Basagni et al., 2001], the shared session key between any two nodes must be derived [Asokan et al., 2000, Yi et al., 2002, Carman et al., 2000]. Some protocols use a trusted third party to deliver keys to every node [Yi et al., 2003], while other protocols pre-distribute communication keys to all nodes]. [Chan et al., 2003] Since WSNs are self-organized, and trusted third party may not be available, key pre-distribution protocols are often adopted in such networks. However, key pre-distribution protocols need to store session keys in every node. This may be difficult to achieve in a sensor network where thousands of nodes are deployed with limited storage space only enough to store a small number of session keys. It is desirable to design a new key pre-distribution protocol, which can reduce the storage space of session keys for a large WSN without degrading its security.

Much research has been done on key distribution in WSN over the past few years. Carman et al. [Carman et al., 2002] analyzed various conventional approaches for key generation and key distribution in WSN on different hardware platforms with respect to computation overhead and energy consumption [Hodjat et al., 2002, Heinelman et al., 2000]. The results showed that conventional key generation and distribution protocols are not suitable for WSN. To cope with the problem, a key management protocol [Carman et al., 2002] is proposed for sensor networks, which is based on group key agreement protocols and identity-based cryptography . This protocol used Diffie-Hellman key exchange scheme to perform group key agreement . However, the high storage and high computation requirements make it difficult to use.

Perrig et al. [Perrig et al., 2001] proposed a security protocol for sensor networks named SPINS . SPINS uses base station as a trusted third party to set up session keys between sensor nodes. Liu and Ning [Liu et al., 2003] extended Perrig's scheme and proposed an efficient broadcast authentication method for sensor networks. Their scheme uses multi-level key chains to distribute the key chain commitments for the broadcast authentication. Undercoffer et al. [Undercoffer et al., 2002] proposed a resource-driven security protocol , which consider the trade-off between security levels and computational resources. However, in a randomly dispersed wireless sensor network, the base station is not always available for all nodes. Without the base station, a sensor network using SPINS may be disconnected. Therefore, these schemes are not well suitable for sensor networks due to the need of base station. Eschenauer and Gligor [Eschenauer et at., 2002] proposed a key management scheme based on Random Graph Theory . [Chan et al., 2003, Erdoos et al., 1960, Spencer, 2000] The Random Graph Theory is defined as follows. A random graph $G(n, p)$ is a graph with n nodes, and the probability that a link exists between any two nodes in the graph is p. When p is equal to 0, the graph G has no edges, whereas when p is equal to 1, the graph G is fully connected. Erdõs and Rẽnyi [Erdoos et al., 1960] showed the monotone properties of a random graph $G(n, p)$ that there exists a threshold value of p, over which value the property exhibits a "phase transition", i.e. the probability for G to have that property will transit from "likely false" to "likely true". The threshold probability is defined by:

$$p = \frac{ln(n) - ln(-ln(P_c))}{n} \quad (7.1)$$

where P_c stands for desired probability of the property. Furthermore, the expected degree of a node can be calculated by:

$$d = p * (n-1) = \frac{(n-1)ln(n) - ln(-ln(P_c))}{n} \quad (7.2)$$

Therefore, the scheme only needs to select d keys to keep a network connected under probability p. It can then significantly reduce the key space. However, it is discovered that the degree d is proportional to the number of nodes n under the same connectivity probability p. That is, when more nodes are deployed, more storage space is needed in each sensor node. Since the storage space in each node is fixed, the maximum number of nodes that can be deployed is also fixed in this scheme. This characteristic restricts the deployment of sensor nodes and therefore the scalability of this scheme is somewhat limited. To improve the scalability, we propose two key distribution schemes: Adaptive Random Pre-distributed scheme (ARP) and Uniquely Assigned One-way Hash Function scheme (UAO) . Both ARP and UAO schemes pre-distribute keys in each sensor node before its deployment. When the number of sensor nodes increases, both key distribution schemes dynamically adjust itself according to remaining storage space in each sensor node without reducing the connectivity probability p. Both schemes minimize the storage requirement for key management under the same connectivity probability p, and can work well even when a large number of sensor nodes are deployed. In contrast, ARP scheme needs the smallest storage space, while UAO scheme provides the capability of

The rest of this paper is organized as follows: The Adaptive Random Pre-distributed scheme and the Uniquely Assigned One-way Hash Function scheme are presented in Sections II and III, respectively. The evaluation of the schemes are provided in Section IV. Finally, Section V concludes the paper.

2. Adaptive Random Pre-distribution Scheme

ARP scheme is composed of two parts. One is the key pool, and the other is the key selection algorithm . The key pool is used to store randomly generated keys, and the key selection algorithm is to select a set of keys from the key pool. Every node needs to select a set of keys from the key pool by using key selection algorithm before its deployment. These selected keys are saved in each node's storage space. Any two nodes shares a common key is able to securely communicate with each other by using this shared key. In ARP, the key pool is a two-dimensional key pool in which keys are generated in two phases, and are arranged in two-dimensional ordered matrix. The key is pre-generated as follows:

2.1 Key Pool Generation Algorithm

- **Step 1:** Randomly generate t keys, called seed keys, and any t one-way hash functions.
- **Step 2:** For every seed key and one-way hash function, an one-way key chain is generated.

It uses $K_{i,0}$ as initial input, and computes the generated key with an one-way hash function F_i . The generated key is fed back into F_i to generate a third key. The procedure $K_{i,j+1} = F(K_{i,j})$ is repeated until the entire key chain is generated.

Consequently, the key chain KC_0 of length s, is composed of a series of keys, $K_{i,0}$, $K_{i,1}$, ..., $K_{i,s-1}$. With t seed keys and t one-way hash functions, t key chains generated, namely KC_0, KC_1,..., KC_{t-1} .

Figure 7.1 demonstrates the difference between the conventional random key pool and the Two-Dimension Key Pool. As shown in Figure 7.1(a), the original random key pool can be regarded as a set of keys disorderly spread into a large pool. In Figure 7.1(b), keys of the Two-Dimension Key Pool are arranged in an s by t matrix.

2.2 Key Selection Algorithm

After key generation, a key pool of size st is generated. Each sensor needs to randomly choose keys from the key pool by using key selection algorithm described here. The number of keys can affect the connectivity of the entire sensor network and the storage requirement of each sensor node. Fewer keys

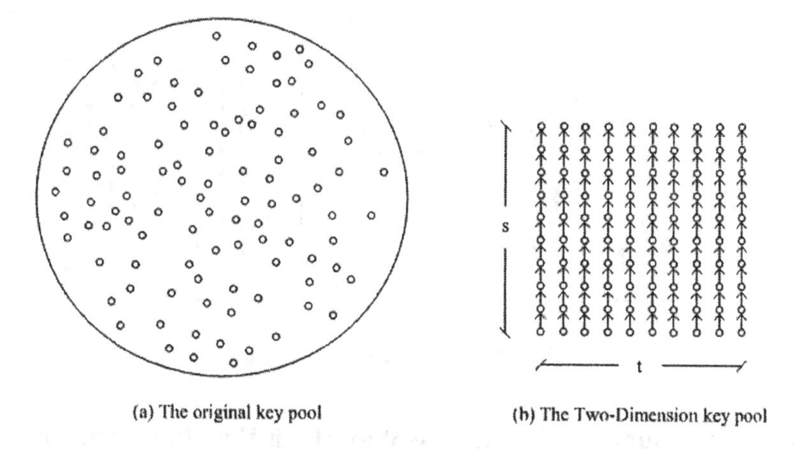

Figure 7.1. Unordered key pool and the Two-Dimension key pool with $t = 10$, $s = 10$.

can save storage but lower the probability for a sensor network to be connected. More keys can guarantee higher connectivity probability but at the same time increase the storage requirement. We'll discuss the the relationship of keys, connectivity probability and storage requirement later in the paper.

The key selection algorithm is used to select a set of communication keys by all nodes before its deployment. The detail of the key selection algorithm for ARP scheme is described as follows.

- Step1: Let r be the number of keys each node needed to achieve connectivity among n sensor node with probability p. r can be chosen as d in eq.2. Each sensor node randomly picks up an one-way key chain $KC_i = (KC_{i,0}, KC_{i,1}, \ldots, KC_{i,s-1})$ from the two-Dimension key pool, and use the keys in the key chain.

- Step2: Each sensor node randomly selects the remaining $r' = r - s$ keys from different key chains.

- Step3: Each sensor node has chosen one key chain KC_i and r' single keys. For each sensor node, it will only need to memorize those r' keys and the one-way hash function F_i and seed key $KC_{i,0}$ of the key chain KC_i.

Figure 7.2 shows an example of key selection, where $t = 10$, $s = 10$, and $r' = 5$. The randomly selected one-way key chain is KC_3, and the rest ri' randomly picked keys are $KC_{0,6}$, $KC_{5,8}$, $KC_{6,3}$, $KC_{8,7}$, and $KC_{9,4}$.

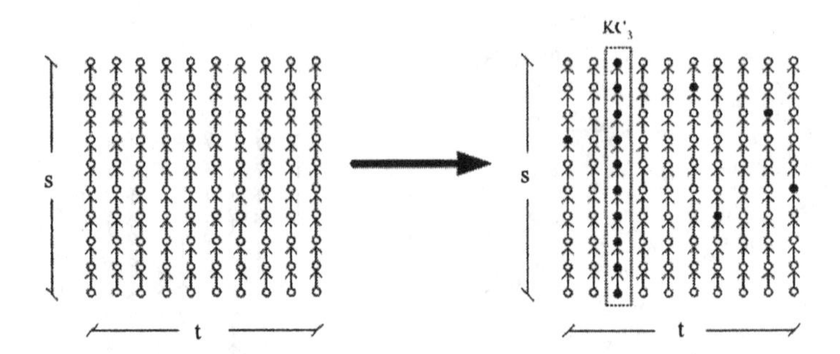

Figure 7.2. A key selection example

3. Uniquely Assigned One-Way Hash Function Scheme

In ARP, any two nodes shared a can directly communicate with each other in a secure way. However, a key in ARP may be shared by more than two nodes, and therefore a node may not be able to authenticate with the shared key the identity of an individual. To cope with the problem, UAO extends ARP to authenticate individual sensor node identities. The detail of UAO is describes as follows.

Each sensor node SN_i is assigned a unique identity ID_i and a uniquely assigned one-way hash function F_i before its deployment. In contrast to ARP key selection algorithm, UAO scheme does not select a key. Instead, it uses ID_i and F_i to decide a key, where ID_i can be the node's MAC address or identifier; and F_i is an one-way hash function. The UAO key decision algorithm is as follows:

3.1 Key Decision Algorithm:

- Step1: Assume the required number of keys to achieve the Random Graph Theory is r. For each sensor node SN_i in the network, SN_i will randomly select r sensor nodes, excluding itself, in the network, denoted as $SN_{v1}, SN_{v2}, \ldots, SN_{vr}$.

- Step 2: For each sensor node SN_{vj}, where j ranges from 1 to r, SN_{vj} uses its unique one-way hash function F_j to generate a unique K_j for SN_i. The K_j is generated by the following equation:

$$K_j = F_j(ID_i)$$

SN_i will memorize all pairs of K_j and ID_j in its key ring.

3.2 Mutual Authentication

After applying key decision algorithm , every node is deployed in a WSN. For communication between two nodes, SN_i and SN_j, SN_i shares unique session key K_j with SN_j, and SN_j shares unique session key K_i with SN_i. Mutual authentication is achieved because SN_i is the only node that owns the unique one-way hash function F_i. If SNi can correctly calculate K_j and decrypt the cipher, then SN_j can authenticate the identity of SN_i. Due to K_j is derived from F_i and ID_j, if SN_j really owns the key K_j then it will make the correct response. Therefore the SN_i will be able to authenticate SN_j with ID_j.

4. Evaluation

To evaluate ARP scheme and UAO scheme, both schemes are analyzed in terms of connectivity and storage space.

4.1 Evaluation of ARP Scheme

To evaluate ARP scheme, the connectivity probability is analyzed because it was observed in the preceding section that ARP is proposed based on Random Graph Theory. If the connectivity probability of different schemes is the same, the scheme requires smaller storage space to store keys.

To evaluate the required probability of connectivity, the network size n and the expected probability Pc of forming a connected graph must be determined. By given n and P_c, we can calculate the threshold probability p and the expected degree d by Equation 7.1 and 7.2. Moreover, since a sensor node cannot communicate with all other nodes in the network, only a limited number of neighbor nodes n' can be contacted. Therefore, the probability of sharing a common key between any two nodes in a neighborhood is:

$$p' = \frac{d}{n'} \tag{7.3}$$

Also, the required key ring size s and the key pool size K to achieve the probability of neighborhood connectivity can be determined.

We denote the probability of any two nodes in the neighborhood sharing at least one common key in Two-Dimension Key Pool Selecting scheme as p'. It is proved that p' is related to the number of key chains t, key chain length s, and the number of selected keys r'. The p' can be calculated by one minus the probability that any two nodes in the neighborhood do not sharing any key. To calculate the probability that any two nodes A and B do not sharing any key, the calculation can be categorized into four parts:

1 A's one-way key chain does not match with B's one-way key chain.

2 A's one-way key chain does not match with any B's selected keys.

3 A's selected keys do not match with B's one-way key chain.

4 A's selected keys do not match with any B's selected keys.

Since B selects one hash function and r' selected keys in different key chains, A's one-way key chain must belong to the rest of the $t - (r' + 1)$ key chains. Therefore, the probability of matching both the first and the second conditions are $\frac{t-(r'+1)}{t}$.

For the third condition, we randomly choose r' key chains from the key pool. A's r' selected keys must not belong to A's key chain. As to match the third condition, it must not also belong to B's key chain. Thus the probability can be calculated as

$$\frac{\binom{t-2}{r'}}{\binom{t-1}{r'}} = \frac{t-r'-1}{t-1}$$

For the fourth condition, it is assumed that A and B have exactly i selected keys belonging to the same i key chains and the probability that A and B have exactly i selected keys belonging to the same i key chains as $p(i)$. There are $\binom{r'}{i}$ ways to pick i common key chains from B's selected key ring, and there are only $(t - 2 - r')$ key chains to pick up the remaining A's $(r' - i)$ selected keys. This is because we have to eliminate A's and B's key chains and the other r' key chains that B's r' selected keys belong to. Thus there are $\binom{t-2-r'}{r'-i}$ ways to pick up the remaining $(r' - i)$ key chains. The total number of ways for A to choose r' key chains is $\binom{t-2}{r'}$. Therefore we get the following equation:

$$p(i) = \frac{\binom{r'}{i}\binom{t-2-r'}{r'-i}}{\binom{t-2}{r'}}$$

Moreover, considering that A and B have exactly i selected keys belonging to the same key chains, the probability that A's selected keys do not match with any B's selected keys becomes:

$$p(i)(1 - \frac{1}{s})^t$$

Hence, to calculate the probability of matching the fourth condition, we have to consider all possible value of i, where $i = 0, 1, 2, \ldots, r'$. Thus the probability for the fourth condition is:

$$\sum_{i=0}^{r'} p(i)(1 - \frac{1}{s})^t$$

By Summarizing the above four conditions, we can calculate the probability p' by the following equation:

$$p' = 1 - \left(\frac{t - (r' + 1)}{t}\right) \left(\frac{t - r' - 1}{t - 1}\right) \left(\sum_{i=0}^{r'} p(i)(1 - \frac{1}{s})^t\right)$$

Figure 7.3 shows the probability of connectivity with different configurations of number of key chains t and the key chain length s.

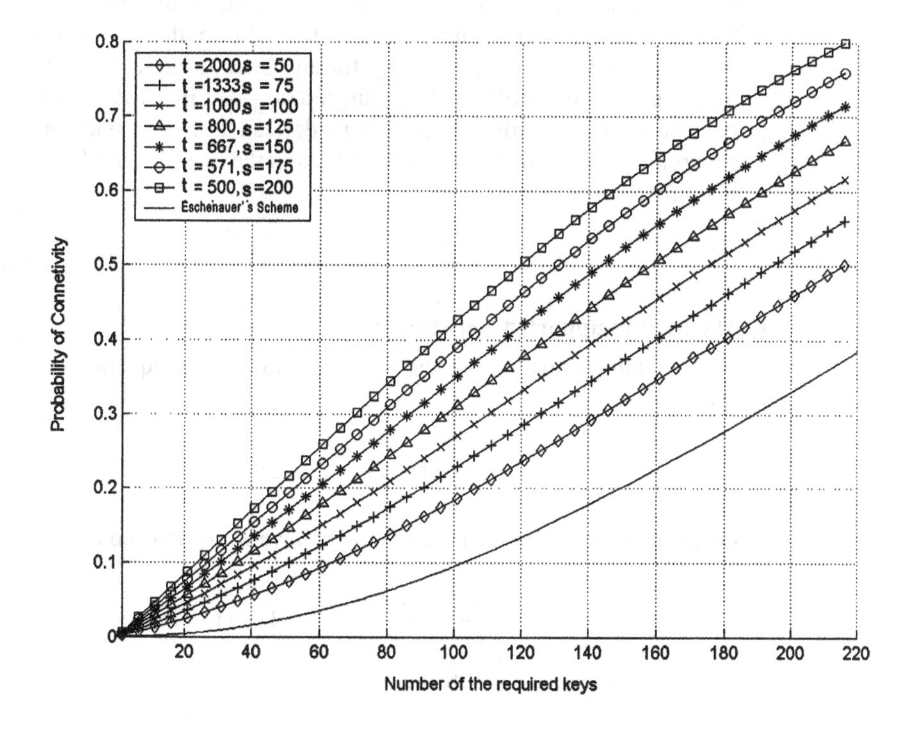

Figure 7.3. Comparison of different configured Two-Dimension Key Pool Selecting Schemes and Eschenauer's scheme (key pool size is 100,000)

As Figure 3 shows, under the same connectivity probability, the ARP scheme requires fewer keys than Eschenauer's scheme . In other words, the ARP scheme demands for less storage space than the Eschenauer's scheme does. Moreover, with different h and y value, the ARP scheme needs different storage space. This can be left as an option for deployment consideration.

4.2 Evaluation of UAO

In this section, evaluation of the probability of connectivity and the maximum supported network size are analyzed. The maximum supported network size stands for maximum sensor node capacity that can achieve mutual authentication under the same memory storage space attached in every sensor node. In addition, we also make a comparison with the random-pairwise scheme in terms of maximum supported network size and the probability of connectivity.

- **Probability of Connectivity:**

  In UAO scheme, the probability of any two nodes in the neighborhood sharing a common key can be evaluated by one minus the probability of that either nodes does not have any key derived from the other's unique one-way function. The probability for any node to get a key derived from a particular node's one-way function is $\frac{r}{n-1}$. Because each node gets r keys in the key ring, those keys are derived from other nodes in the network. The probability of any two nodes in the neighborhood sharing a common key will be

$$p' = 1 - (1 - \frac{r}{n-1})^2 \tag{7.4}$$

- **Maximum Supported Network Size:**

  By combining Equation 7.3 and 7.4, the following the equation can be derived.

$$\frac{d}{n'} = 1 - (1 - \frac{r}{n-1})^2$$

  Furthermore, by using Equation 7.2, the above equation becomes:

$$\frac{(n-1)(ln(n) - ln(-ln(P_c)))}{n * n'} = 1 - (1 - \frac{r}{n-1})^2$$

  The equation can be simplified to:

$$r^2 - 2(n-1)r + (n-1)^2 \frac{(n-1)(ln(n) - ln(-ln(P_c)))}{n * n'} = 0$$

By calculating the root of the above quadratic equation, we can get:

$$r = (n-1)(1 - \sqrt{1 - \frac{(n-1)(ln(n) - ln(-ln(P_c)))}{n * n'}}) \quad (7.5)$$

It can be more simplified as:

$$r = (n-1)(1 - \sqrt{1 - \frac{d}{n'}})$$

In comparison with the Random-Pairwise scheme , we assume the network size is n, expected degree of graph connectivity is d, the number of neighbor nodes is n', and the key ring size is r. According to the definition of pairwise scheme, there are only r nodes having common shared keys with each sensor node and it still has to achieve the expected degree in the neighborhood. Then we can find the following equation:

$$d = \frac{r * n'}{n} \Rightarrow r = \frac{d * n}{n'} \quad (7.6)$$

To analyze the relationship between memory space and network size, first we combine Equation 7.2 and Equation 7.6 to obtain the following equation:

$$r = \frac{(n-1)}{n'}(ln(n) - ln(-ln(P_c))) \quad (7.7)$$

According to the Equation 7.7, it is clear that the complexity of memory space requirement for the Random-Pairwise scheme is $O(nln(n))$. In addition, according to the Equation 7.5, it is found that the complexity of memory space requirement for the UAO scheme is $O(n\sqrt{ln(n)})$.

Figure 4 shows the comparisons of UAO scheme and Random-Pairwise keys distribution scheme in memory space requirement and the maximum supported network size. As Figure 4 shows, UAO scheme achieves better performance in maximizing network size under the same memory requirement. Therefore, with the same sensor node hardware equipment, UAO can adapt more sensor nodes in a network while providing better security than Random-Pairwise key distribution scheme.

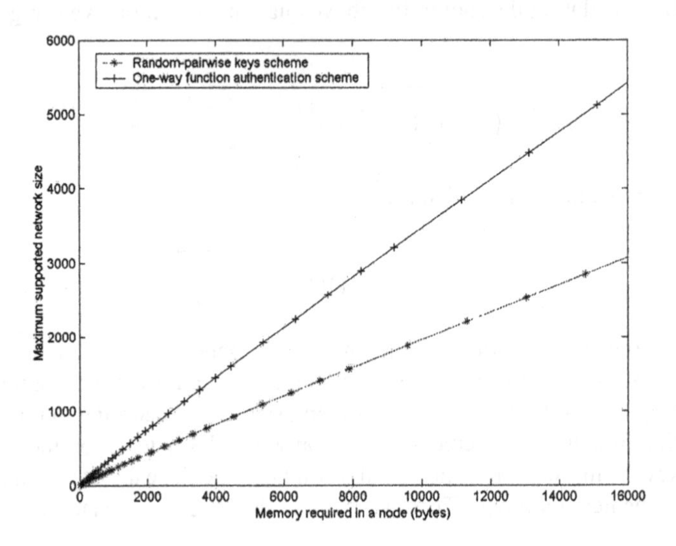

Figure 7.4. Comparison of Random-pairwise keys scheme and UAO scheme in memory requirement and maximum supported network size.

5. Conclusion

Key distribution is a critical and fundamental issue for the security service in wireless sensor networks. The pre-distributed and symmetric cryptography based key management system is well suitable for the resource limited sensor network. Two efficient schemes are proposed which are based on the Random Graph Theory to provide key distribution for the secure sensor network services.

Adaptive Random Pre-distributed scheme needs less memory space than existing schemes. ARP can be used in the WSN with a large number of nodes where each node contains limited storage space. On the other hand, Uniquely Assigned One-Way Hash Function scheme possesses the characteristics of mutual authentication . The tradeoff between these two schemes depends on security requirement, network size and available memory space. If mutual authentication of individuals is desirable, Uniquely Assigned One-Way Hash Function scheme should be used. Otherwise, the Adaptive Random Pre-distributed scheme should be used because it needs smaller storage space.

References

[Akyilidiz et al., 2002] F. Akyildiz, W. Su, Y. Sankarasubramaniam, and E. Cayirci. (2002). A Survey on Sensor Networks. In *IEEE Communications Magazine*, pages 102-114, August.

[Estrin et al., 1999] D. Estrin, R. Govindan, J. Heidemann and S. Kumar. (1999). Next Century Challenges: Scalable Coordination in Sensor Networks, In *Proceedings of the Fifth Annual ACM/IEEE International Conference on Mobile Computing and Networking*, August.

[Slijepcevic et al., 2002] S. Slijepcevic, M. Potkonjak, V. Tsiatsis, S. Zimbeck, and M. B. Srivastava. (2002). On communication security in wireless ad-hoc sensor network, *Eleventh IEEE International Workshops on Enabling Technologies: Infrastructure for Collaborative Enterprises (WETICE'02)*, June.

[Tilak et al., 2002] S. Tilak, N. Abu-Ghazaleh, and W. Heinzelman. (2002). A taxonomy of wireless microsensor network models, *ACM Mobile Computing and Communications Review (MC2R 2002)*.

[Hodjat et al., 2002] Hodjat and I. Verbauwhede. (2002). The energy cost of secrets in adhoc networks, *IEEE CAS Workshop on Wireless Communications and Networking*, September.

[He et al., 2003] T. He, J. A. Stankovic, C. Lu, and T. F. Abdelzaher. (2003). Speed: A stateless protocol for real-time communication in sensor networks, In *International Conference on Distributed Computing Systems (ICDCS 2003)*, May.

[Heinelman et al., 2000] W. Heinzelman, A. Chandrakasan, and H. Balakrishnan. (2000). Energy efficient communication protocols for wireless microsensor networks, *Proc. Hawaaian Int'l Conf. on Systems Science*, January.

[Heinzelman et al., 1999] W. Heinzelman, J. Kulik, and H. Balakrishnan. (1999). Adaptive protocols for information dissemination in wireless sensor networks, In *Proceedings of the Fifth Annual ACM/IEEE International Conference on Mobile Computing and Networking*, August.

[Intanagonwiwat et al., 2000] C. Intanagonwiwat, R. Govindan, and D. Estrin. (2000). Directed diffusion: A scalable and robust communication paradigm for sensor networks, In *Proceedings of the Sixth Annual International Conference on Mobile Computing and Networks (MobiCOM '00)*, August 2000.

[Zhou et al., 1999] L. Zhou and Z. J. Haas. (1999) Securing ad hoc networks, *IEEE Networks Magazine*, vol. 13, no. 6, pages 24-30, November.

[Kong et al., 2001] J. Kong, P. Zerfos, H. Luo, S. Lu, and L. Zhang. (2001). Providing robust and ubiquitous security support for mobil ad-hoc network, *Network Protocols Ninth International Conference on ICNP 2001*.

[Luo et al., 2002] H. Luo, P. Zerfos, J. Kong, S. Lu, and L. Zhang. (2002). Self-securing ad hoc wireless networks, In *Proceedings of Seventh International Symposium on Computers and Communications (ISCC 2002)*, pages 567-574.

[Hubaux et al., 2001] J.-P. Hubaux, L.Buttyan, and S. Capkun. (2001). The quest for security in mobile ad hoc networks, In *Proceedings of the 2001 ACM International Symposium on Mobile Ad Hoc Networking and Computing*, October.

[Asokan et al., 2000] N. Asokan and P. Ginzborg. (2000). in ad hoc networks, *Computer Communications*, vol. 23, pages 1627-1637.

[Yi et al., 2003] S. Yi and R. Kravets. (2003). Moca: Mobile certificate authority for wireless ad hoc networks, *2nd Annual PKI Research Workshop Program (PKI03)*, April.

[Yi et al., 2002] Seung Yi and Robin Kravets. (2002) Key management for heterogeneous ad hoc wireless networks, *The 10th IEEE International Conference on Network Protocols (ICNP2002)*.

[Basagni et al., 2001] S. Basagni, K. Herrin, E. Rosti, D. Bruschi, and E. Rosti. (2001). Secure pebblenets, In *Proceedings of the 2001 ACM International Symposium on Mobile Ad Hoc Networking and Computing*, pages 156 - 163.

[Carman et al., 2000] D. W. Carman, P. S. Kruus, and B. J. Matt. (2000). Constraints and approaches for distributed sensor network security, *NAI Labs Technical Report #00- 010*, September.

[Carman et al., 2002] D. W. Carman, B. J. Matt, and G. H. Cirincione. (2002). Energy-efficient and low-latency key management for sensor networks, In Proceedings of 23rd Army Science Conference, December 2002.

[Perrig et al., 2001] A. Perrig, R. Szewczyk, V. Wen, D. Culler, and J. D. Tygar, Spins: Security protocols for sensor networks, In *Proceedings of the Seventh Annual International Conference on Mobile Computing and Networking*, July.

[Liu et al., 2003] D. Liu and P. Ning. (2003). Efficient distribution of key chain commitments for broadcast authentication in distributed sensor networks, *The 10th Annual Network and Distributed System Security Symposium*, February.

[Undercoffer et al., 2002] J. Undercoffer, S. Avancha, A. Joshi, and J. Pinkston. (2002). Security for sensor networks, *2002 CADIP Research Symposium*.

[Eschenauer et at., 2002] L. Eschenauer and V. D. Gligor. (2002). A key-management scheme for distributed sensor networks, In *Proceedings of the 9th ACM Conference on Computer and Communication Security*, pages 41-47, November.

[Chan et al., 2003] H. Chan, A. Perrig, and D. Song. (2003). Random key pre-distribution schemes for sensor networks, *IEEE Symposium on Security and Privacy*, May.

[Erdoos et al., 1960] P. Erdõs and A. Rẽnyi. (1960). On the evolution of random graphs, *Publ. Math. Inst. Hungat. Acad. Sci.*, vol. 5, pages 17-6.

[Spencer, 2000] J. Spencer. (2000). The Strange Logic of Random Graphs. *Springer-Verlag*.

III

INTRUSION DETECTION, DEFENSE, MEASUREMENT

Chapter 8

MEASURING RELATIVE ATTACK SURFACES

Michael Howard
Security Business Unit
Microsoft Corporation
Redmond, WA

Jon Pincus
Microsoft Research
Microsoft Corporation
Redmond, WA

Jeannette M. Wing
School of Computer Science
Carnegie Mellon University
Pittsburgh, PA

Abstract We propose a metric for determining whether one version of a system is more secure than another with respect to a fixed set of dimensions. Rather than count bugs at the code level or count vulnerability reports at the system level, we count a system's *attack opportunities*. We use this count as an indication of the system's "attackability ," likelihood that it will be successfully attacked. We describe a system's *attack surface* along three abstract dimensions: targets and enablers, channels and protocols, and access rights. Intuitively, the more exposed the system's surface, the more attack opportunities, and hence the more likely it will be a target of attack. Thus, one way to improve system security is to reduce its attack surface.

To validate our ideas, we recast Microsoft Security Bulletin MS02-005 using our terminology, and we show how Howard's Relative Attack Surface Quotient for Windows is an instance of our general metric.

Keywords: Security metrics, attacks, vulnerabilities, attack surface, threat modeling.

1. Introduction

Given that security is not an either-or property, how can we determine that a new release of a system is "more secure" than an earlier version? What metrics should we use and what things should we count? Our work argues that rather than attempt to measure the security of a system in absolute terms with respect to a yardstick, a more useful approach is to measure its "relative" security. We use "relative" in the following sense: Given System A, we compare its security *relative to* System B, and we do this comparison with respect to a given number of yardsticks, which we call *dimensions*. So rather than say "System A is secure" or "System A has a measured security number N" we say "System A is more secure than System B with respect to a fixed set of dimensions."

In what follows, we assume that System A and System B have the same operating environment. That is, the set of assumptions about the environment in which System A and System B is deployed is the same; in particular, the threat models for System A and System B are the same. Thus, it helps to think of System A and System B as different versions of the same system.

1.1 Motivation

Our work is motivated by the practical problem faced in industry today. Industry has responded to demands for improvement in software and systems security by increasing effort[1] into creating "more secure" products and services. How can industry determine if this effort is paying off and how can we as consumers determine if industry's effort has made a difference?

Our approach to measuring relative security between systems is inspired by Howard's informal notion of *relative attack surface* [Howard, 2003]. Howard identified 17 "attack vectors," i.e., likely opportunities of attack. Examples of his attack vectors are open sockets, weak ACLs, dynamic web pages, and enabled guest accounts. Based on these 17 attack vectors, he computes a "measure" of the attack surface, which he calls the Relative Attack Surface Quotient (RASQ) , for seven running versions of Windows.

We added three attack vectors to Howard's original 17 and show the RASQ calcuation for five versions of Windows in Figure 8.1. The bar chart suggests that a default running version of Windows Server 2003 is much more secure than previous versions with respect to the 20 attack vectors. It also illustrates that the attack surface of Windows Server 2003 increases only marginally when IIS is enabled—in sharp contrast to Windows NT 4.0, where enabling IIS (by installing the "Option Pack") dramatically increased the RASQ, and to Windows 2000, where IIS is enabled by default[2]. As will be discussed in Section 6.3, these differences in RASQ are consistent with anecdotal evidence for the relative security of different Windows platforms and configurations.

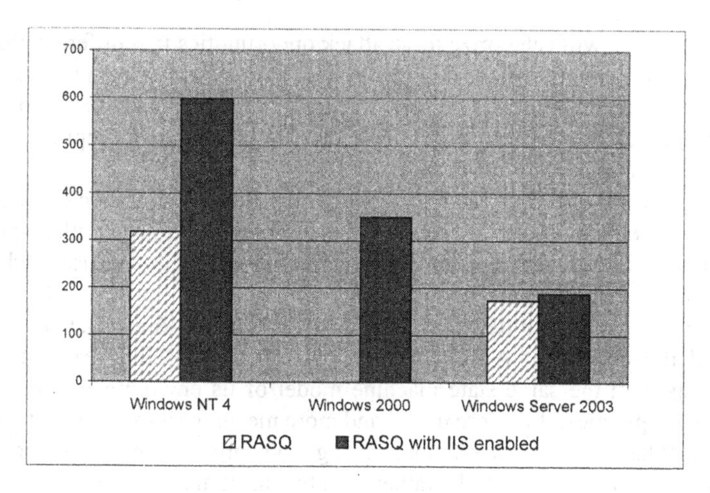

Figure 8.1. Relative Attack Surface Quotient of Different Versions of Windows [Howard, 2003]

1.2 A New Metric: Attackability

Two measurements are often used to determine the security of a system: at the code level, a count of the number of bugs found (or fixed from one version to the next); and at the system level, a count of the number of times a system (or any of its versions) is mentioned in the set of Common Vulnerabilities and Exposures (CVE) bulletins [CVE], CERT advisories [CERT], etc.

Rather than measure code-level or system-level vulnerability, we consider a different measure, somewhat in between, which we call *attack opportunity*, or "attackability" for short. Counting the number of bugs found (or fixed) misses bugs that are not found (or fixed), perhaps the very one that is exploited; it treats all bugs alike when one might be easier to exploit than another, or the exploit of one may result in more damage than the exploit of another. Instead, we want a measure—at a higher abstraction level—that gives more weight to bugs that are more likely to be exploited. Counting the number of times a system version appears in bulletins and advisories ignores the specifics of the system configuration that give rise to the exploit: whether a security patch has been installed, whether defaults are turned off, whether it always runs in system administrator mode. Instead, we want a measure—at a lower abstraction level—that allows us to refer to very specific states (i.e., configurations) of a system. Given this intermediate viewpoint, we propose that there are certain system features that are more likely than others to be opportunities of attack. The counts of these "more likely to be attacked" system features determine a system's attackability.

Further, we will categorize these attack opportunities into different abstract *dimensions*, which together define a system's *attack surface*. Intuitively, the more exposed the system's surface, the more attack opportunities, and hence the more likely it will be a target of attack. Thus, one way to improve system security is to reduce its attack surface.

Suppose now we are given a fixed set of dimensions and a fixed set of attack opportunities (i.e., system features) for each dimension. Then with respect to this fixed set of dimensions of attack opportunities, we can measure whether System A is "more secure" than System B.

In our work, we use state machines to model Systems A and B. Our abstract model allows Systems A and B to be any two state machines, each of which interacts with the same state machine model of its environment, i.e., threat model. In practice, it is more useful and more meaningful to compare two systems that have some close relationship, e.g., they provide similar functionality, perhaps through similar APIs, rather than two arbitrary systems. The abstract dimensions along which we compare two systems are derived directly from our state machine model: *process and data resources* and the *actions* that we can execute on these resources. For a given attack, which we define to be a sequence of action executions, we distinguish *targets* from *enablers*: *targets* are processes or data resources that an adversary aims to control, and *enablers* are all other processes and data resources that are used by the adversary to carry out the attack successfully. The adversary obtains control over these resources through communication *channels* and *protocols*. Control is subject to the constraints imposed by a system's *set of access rights*. In summary, our attack surface's three dimensions are: targets and enablers, channels and protocols, and access rights. Attackability is a measure of how exposed a system's attack surface is.

1.3 Contributions and Roadmap

We use a state machine formal framework to support three main contributions of this paper:

- The notion of a system's *attack surface.*
- A new *relative measure of security*, attackability.
- A model for vulnerabilities as differences between intended and actual behavior, in terms of pre-conditions and post-conditions (Section 2.2).

Our "relative" approach has the advantage that security analysts are more willing and able to give relative rankings of threats and relative values to risk-mitigation controls, than absolute numbers [Butler, 2003]. We also avoid the need to assign probabilities to attacks.

We view our work as only a first step toward coming up with a meaningful, yet practical way of measuring (relative) security. By no means do we claim to have identified "the right" or "all" the dimensions of an attack surface. Indeed, our use of the word "dimensions" is only meant to be suggestive of a surface; our dimensions are not orthogonal. We hope with this paper to spark a fruitful line of new research in security metrics.

In Section 2 we present our formal framework and then in Section 3 we explain our abstract dimensions of a system's attack surface. To illustrate these ideas concretely, in Section 4 we recast Microsoft Security Bulletin MS02-005 in terms of our concepts of targets and enablers. In Section 5 we give an abstract attack surface measurement function. Again, to be concrete, in Section 6 we revisit Howard's RASQ metric in terms of our abstract dimensions. In Section 7 we discuss how best to apply and not to apply the RASQ approach. We close with a review of related work in Section 8 and suggestions for future work in Section 9.

2. Terminology and Model

Our formal model is guided by the following three terms from *Trust in Cyberspace* [Schneider, 1991]:

- A *vulnerability* is an error or weakness in design, implementation, or operation.
- An *attack* is the means of exploiting a vulnerability.
- A *threat* is an adversary motivated and capable of exploiting a vulnerability.

We model both the system and the threat as state machines, which we will call *System* and *Threat*, respectively. A state machine has a set of states, a set of initial states, a set of actions, and a state transition relation. We model an attack as a sequence of executions of actions that ends in a state that satisfies the adversary's goal, and in which one or more of the actions executed in an attack involves a vulnerability.

2.1 State Machines

A *state machine*, $M = \langle S, I, A, T \rangle$, is a four-tuple where S is a set of states, $I \subseteq S$ is a set of initial states, A is a set of actions, and $T = S \times A \times S$ is a transition relation. A state $s \in S$ is a mapping from typed *resources* to their typed *values*:

$$s : Res_M \rightarrow Val_M$$

Of interest to us are state resources that are *processes* and *data*. A *state transition*, $\langle s, a, s' \rangle$, is the execution of action a in state s resulting in state s'. A change in state means that either a new resource is added to the mapping, a resource was deleted, or a resource changes in value. We assume each state transition is atomic.

An *execution* of a state machine is the alternating sequence of states and action executions:

$$s_0\ a_1\ s_1\ a_2\ s_2\ \ldots\ s_{i-1}\ a_i\ s_i\ \ldots$$

where $s_0 \in I$ and $\forall i > 0. \langle s_{i-1}, a_i, s_i \rangle \in T$. An execution can be finite or infinite. If finite, it ends in a state.

The *behavior* of a state machine, M, is the set of all its executions. We denote this set *Beh(M)*. A state s is *reachable* if either $s \in I$ or there is an execution, $e \in Beh(M)$, such that s appears in e.

We will assume that actions are specified by pre- and post-conditions. For an action, $a \in A$, if *a.pre* and *a.post* denote a's pre- and post-condition specifications, we can then define the subset of the transition relation, T, that involves only action a as follows:

$$a.T = \{\langle s, a, s' \rangle : S \times A \times S \mid a.pre(s) \Rightarrow a.post(s, s')\}$$

We model both the system under attack and the threat (adversary) as state machines:

$$System = \langle S_S, I_S, A_S, T_S \rangle$$
$$Threat = \langle S_T, I_T, A_T, T_T \rangle$$

We partition the resources of a state machine, M, into a set of *local resources* and a set of *global resources*, $Res_M = Res_M^L \uplus Res_M^G$. We define the *combination* of the two state machines, *ST* = *System* ⋈ *Threat*, by merging all the corresponding components[3]:

- $S_{ST} \subseteq 2^{Res_{ST} \rightarrow Val_{ST}}$
- $I_{ST} = I_S \cup I_T$
- $A_{ST} = A_S \cup A_T$
- $T_{ST} = T_S \cup T_T$

We identify the global resources of S and the global resources of T such that $Res_{ST}^G = Res_S^G = Res_T^G$ and so $Res_{ST} = Res_S^L \uplus Res_T^L \uplus Res_{ST}^G$. Finally, $Val_{ST} = Val_S \cup Val_T$. We extend the definitions of executions, behaviors, etc. in the standard way.

An adversary targets a system under attack to accomplish a goal:

$$System\text{-}Under\text{-}Attack = (System \bowtie Threat) \times Goal$$

where *Goal* is formulated as a predicate over states in S_{ST}. Note that we make explicit the goal of the adversary in our model of a system under attack. Example goals might be "Obtain root access on host H" or "Deface website on server S." In other contexts, such as fault-tolerant computing, *Threat* is synonymous with the system's "environment." Thus, we use *Threat* to model environmental failures, due to benign or malicious actions, that affect a system's state.

Intuitively, the way to reduce the attack surface is to ensure that the behavior of *System* prohibits *Threat* from achieving its *Goal*.

2.2 Vulnerabilities

Vulnerabilities can be found at different levels of a system: implementation, design, operational, etc. They all share the common intuition that something in the actual behavior of the system deviates from the intended behavior. We can capture this intuition more formally by comparing the difference between the behaviors of two state machines. Suppose there is a state machine that models the intended behavior, and one that models the actual behavior:

$$Intend = \langle S_{Int}, I_{Int}, A_{Int}, T_{Int} \rangle$$
$$Actual = \langle S_{Act}, I_{Act}, A_{Act}, T_{Act} \rangle$$

We define the vulnerability difference set, *Vul*, to be the difference in behaviors of the two machines:

$$Vul = Beh(Actual) - Beh(Intend)$$

An execution sequence in *Vul* arises from one or more differences between some component of the state machine *Actual* and the corresponding component of *Intend*, i.e., differences between the corresponding sets of (1) states (or more relevantly, reachable states), (2) initial states, (3) actions, or (4) transition relations. We refer to any one of these kinds of differences as a *vulnerability*. Let's consider each of these cases:

1. $S_{Act} - S_{Int} \neq \emptyset$

If there is a difference in state sets then there are some states that are defined for *Actual* that are not intended to be defined for *Intend*. The difference may be due to (1) a resource that is in a state in *Actual*, but not in *Intend* or (2) a value allowed for resource in *Actual* that is not allowed for that resource in *Intend*. (A resource that is not in a state in *Actual*, but is in *Intend* is ok.) The difference may not be too serious if the states in the difference are not reachable by some transition in T_{Act}. If they are reachable, then the difference in transition relations will pick up on this vulnerability. However, even if they are not reachable, it means that if any of the specifications for actions changes in the future, we must be careful to make sure that the set of reachable states in *Actual* is a subset of that of *Intend*.

2. $I_{Act} - I_{Int} \neq \varnothing$

If there is a difference in initial state sets then there is at least one state in which we can start an execution when we ought not to. This situation can arise if resources are not initialized when they should be, they are given incorrect initial values, or when there are resources in an initial actual state but not in any initial intended state.

3. $A_{Act} - A_{Int} \neq \varnothing$

If there is a difference in action sets then there are some actions that can be actually done that are not intended. These actions will surely lead to unexpected behavior. The difference will show up in the differences in the state transition relations (see below).

4. $T_{Act} - T_{Int} \neq \varnothing$

If there is a difference in state transition sets then there is at least one state transition allowed in *Actual* that should not be allowed according to *Intend*. This situation can arise because either (i) the action sets are different or (ii) the pre-/post-conditions for an action common to both action sets are different.

More precisely, for case (ii) where $A_{Act} = A_{Int}$, consider a given action $a \in A_{Int}$. If $a.T_{Act} - a.T_{Int}$ is non-empty then there are some states either in which we can execute a in *Actual* and not in *Intend* or which we can reach as a result of executing a in *Actual* and not in *Intend*. Let $a_{Act}.pre$ and $a_{Int}.pre$ be the pre-conditions for a in *Actual* and *Intend*, respectively, and similarly for their post-conditions. In terms of pre- and post-conditions, *no difference* can arise if

- $a_{Act}.pre \Rightarrow a_{Int}.pre$ and
- $a_{Act}.post \Rightarrow a_{Int}.post$.

Intuitively, if the "actual" behavior is stronger than the "intended" then we are safe.

Given that *Actual* models the actual behavior of the system, then our system combined with the *Threat* machine looks like:

$$\textit{System-Under-Attack} = (\textit{Actual} \bowtie \textit{Threat}) \times \textit{Goal}$$

as opposed to

$$\textit{System-Under-Attack} = (\textit{Intend} \bowtie \textit{Threat}) \times \textit{Goal}$$

again with the expectation that were *Intend* implemented correctly, *Goal* would not be achievable.

In this paper we focus our attention at implementation-level vulnerabilities, in particular, differences that can be blamed on an action's pre-condition or post-condition that is too weak or incorrect. A typical example is in handling a buffer overrun . Here is the intended behavior, for a given input string, s:

length(s) $\leq$ 512 $\Rightarrow$ "process normally" $\wedge$ length(s) $>$ 512 $\Rightarrow$ "report error and terminate"

If the programmer forgot to check the length of the input, the actual behavior might instead be

length(s) $\leq$ 512 $\Rightarrow$ "process normally" $\wedge$ length(s) $>$ 512 $\Rightarrow$ "execute extracted payload"

Here "execute extracted payload" presumably has an observable unintended side effect that differs from just reporting an error.

2.3 Attacks

An attack is the "means of exploiting a vulnerability" [Schneider, 1991]. We model an attack to be a sequence of action executions, at least one of which involves a vulnerability. More precisely, an attack, k, either starts in an unintended initial state or reaches an unintended state through one of the actions executed in k. In general, an attack will include the execution of actions from both state machines, *System* and *Threat*.

The difference between an arbitrary sequence of action executions and an attack is that an attack includes either (or both) (1) the execution of an action whose behavior deviates from the intended (see previous section) or (2) the execution of an action, $a \in A_{Act} - A_{Int}(\neq \emptyset)$. In this second case, the set of unintended behaviors will include behaviors not in the set of intended behaviors since $A_{Act} \neq A_{Int}$.

For a given attack, k, the *means of an attack* is the set of all actions in k and the set of all process and data resources accessed in performing each action in k. These resources include all global and local resources accessed by each action in k and all parameters passed in as arguments or returned as a result to each action executed in k.

3. Dimensions of an Attack Surface

We consider three broad dimensions to our attack surface:

- *Targets and enablers.* To achieve his goal, the adversary has in mind one or more targets on the system to attack. An *attack target,* or simply *target,* is a distinguished process or data resource on *System* that plays a critical role in the adversary's achieving his goal. We use the term *enabler* for any accessed process or data resource that is used as part of the means of the attack but is not singled out to be a target.

- *Channels and protocols. Communication channels* are the means by which the adversary gains access to the targets on *System*. We allow both message-passing and shared-memory channels. *Protocols* determine the rules of interaction among the parties communicating on a channel.

- *Access rights.* These rights are associated with each process and data resource of a state machine.

Intuitively, the more targets, the larger the attack surface . The more channels, the larger the attack surface. The more generous the access rights, the larger the attack surface.

We now look at each of these dimensions in turn.

3.1 Targets and Enablers

Targets and enablers are resources that an attacker can use or coopt. There are two kinds: processes and data. Since it is a matter of the adversary's goal that determines whether a resource is a target or enabler, for the remainder of this section we use the term targets to stand for both. In particular, a target in one attack might simply be an enabler for a different attack, and vice versa.

Examples of process targets are browsers, mailers, and database servers. Examples of data targets are files, directories, registries, and access rights.

The adversary wants to control the target: modify it, gain access to it, or destroy it. Control means more than ownership; more generally, the adversary can use it, e.g., to trigger the next step in the attack. Consider a typical worm or virus attack, which follows this general pattern:

Step 1: Ship an executable—treated as a piece of data—within a carrier to a target machine.

Step 2: Use an enabler, e.g., a browser, to extract the payload (the executable) from the carrier.

Step 3: Get an interpreter to execute the executable to cause a state change on the target machine.

where the attacker's goal, achieved after the third step, may be to modify state on the target machine, to use up its resources, or to set it up for further attacks.

The prevalence of this type of attack leads us to name two special types of data resources. First, *executables* is a distinguished type of data resource in that they can be interpreted (i.e., evaluated). We associate with executables one or more eval functions, *eval: executable* $\rightarrow$ *unit.*[4] Different *eval* functions might interpret the same executable with differing effects. Executables can be targets and controlling such a target includes the ability to call an *eval* function on it. The adversary would do so, for example, for the side effect of establishing the pre-condition of the next step in the attack.

Obvious example types of *eval* functions include browsers, mailers, applications, and services (e.g., Web servers, databases, scripting engines). Less obvious examples include application extensions (e.g., Web handlers, add-on dll's, ActiveX controls, ISAPI filters , device drivers), which run in the same process as the application; and helper applications (e.g., CGI scripts), which run in a separate process from the application.

Carriers are our second distinguished type of data resource. Executables are embedded in carriers. Specifically, carriers have a function *extract_payload: carrier → executable*. Examples of carriers include viruses, worms, Trojan horses, and email messages.

Part of calculating the attack surface is determining the types and numbers of instances of potential process targets and data targets, the types and numbers of instances of eval functions for executables that could have potentially damaging side effects; and the types and numbers of instances of carriers for any executable.

3.2 Channels and Protocols

A channel is a means of communicating information from a sender to a receiver (e.g., from an attacker to a target machine). We consider two kinds of channels: message-passing (e.g., sockets, RPC connections , and named pipes) and shared-memory (e.g., files, directories, and registries). Channel "endpoints" are processes.

Associated with each kind of channel is a *protocol*, the rules of exchanging information. For message-passing channels, example protocols include ftp, RPC, http, and streaming. For shared-memory, examples include protocols that might govern the order of operations (e.g., a file has to be open before read), constrain simultaneous access (e.g., multiple-reader/single-writer or single-reader/single-writer), or prescribe locking rules (e.g., acquire locks according to a given partial order).

Channels are data resources. A channel shared between *System* and *Threat* machines is an element of Res^G_{ST} in the combination of the two machines. In practice, in an attack sequence, the *Threat* machine might establish a new message-passing channel, e.g., after scanning host machines to find out what services are running on port 80.

Part of calculating the attack surface is determining the types of channels, the numbers of instances of each channel type, the types of protocols allowed per channel type, the numbers and types of processes at the channel endpoints, the access rights (see below) associated with the channels and their endpoints, etc.

3.3 Access rights

We associate *access rights* with all resources. For example, for data that are text files, we might associate read and write rights; for executables, we might associate execute rights. Note that we associate rights not only with files and directories, but also with channels (since they are data resources) and channel endpoints (since they are running processes).

Conceptually we model these rights as a relation, suggestive of Lampson's orginal access control matrix [Jampson, 1974]:

$$Access \subseteq Principals \times Res \times Rights$$

where *Principals* = *Users* ∪ *Processes*, *Res* = *Processes* ∪ *Data*, and *Rights* is left uninterpreted. (*Res* is the same set of resources introduced in Section 2.) For example, in Unix, *Rights* = {read, write, execute}, in the Andrew file system, *Rights* = {read, lookup, insert, delete, write, lock, administer}, and in Windows there are eighteen different rights associated with files and directories alone; and of course not all rights are appropriate for all principals or resources. More generally, to represent conditional access rights, we can extend the above relation with a fourth dimension, $Access \subseteq Principals \times Res \times Rights \times Conditions$, where *Conditions* is a set of state predicates.

There are shorthands for some "interesting" subsets of the *Access* relation, e.g., accounts, trust relationships, and privilege levels, that we usually implement in practice, in lieu of representing the *Access* relation as a matrix.

- *Accounts* represent principals, i.e., users and processes. Thus, we view an account as shorthand for a particular principal with a particular set of access rights. Accounts can be data or process targets.

  There are some special accounts that have default access rights. Examples are well-known accounts such as guest accounts, and accounts with "admin" privileges. These typically have names that are easy to guess.

  Part of calculating the attack surface is determining the number of accounts, the number of accounts with admin privileges, and the existence and number of guest accounts, etc. Also, part of calculating the attack surface is determining for each account if the tightest access rights possible are associated with it.

- A *trust relationship* is just a shorthand for an expanded access rights matrix. For example, we might define a specific trust relation, $Tr \subseteq Principals \times Principals$, where network hosts might be a subset of *Principals*. Then we might define the access rights for principal p_1 to be the same as or a subset of those for principal p_2 if $Tr(p_1, p_2)$. We could do something similar to represent the "speaks for" relation of Lampson, Abadi, Burrows, and Wobber [Lampson et al., 1992]. In both cases, by modeling access rights as a (flat) ternary relation, however, we lose some information: the structural relationship between the two principals (A trusts B or A speaks for B). We choose, however, to stick to the simpler access rights matrix model because of its prevalence in use.

- *Privilege levels* map a principal to a level in a total or partial order, e.g., *none* < *user* < *root*. Associated with a given level is a set of access rights. Suppose we have a function, *priv_level*: *Principals* → {none, user, root}, then the rights of principal p would be those associated with *priv_level(p)*.

Reducing the attack surface with respect to access rights is a special case of abiding by the Principle of Least Privilege: Grant only the relevant rights to each of the principals who are allowed access to a given resource.

4. Security Bulletins

To validate our general attack surface model, we described a dozen Microsoft Security bulletins [MSRC] using our terminology [Pincus and Wing, 2003]. The one example we present here illustrates how two different attacks can exploit the same vulnerability via different channels.

The Microsoft Security Bulletin MS02-005, posted February 11, 2002, reports six vulnerabilities and a cumulative patch to fix all of them. We explain just the first (see Figures 8.2 and 8.3). The problem is that the processing of an HTML document (a web page sent back from a server or HTML email) that embeds another object involves a buffer overrun . Exploiting this buffer overrun vulnerability lets the adversary run arbitrary code in the security context of the user.

We now walk through the template which we use for describing these bulletins.

First we specify the vulnerability as the difference in actual from intended behavior for an action. Here the action is the processing by MSHTML (the HTML renderer on Microsoft Windows 2000 and Windows XP) of an HTML document D in a security zone Z. The intended pre-condition is "true," i.e., this action should be allowed in all possible states. However, due to a missing validation check of the action's input, the actual pre-condition is that the length of the object, X, embedded in D, should be less than or equal to 512 bytes.

The intended post-condition is to display the embedded object as long as the ability to run ActiveX Controls is enabled for zone Z. The actual post-condition, due to the non-trivial pre-condition, is that if the length of X is longer than 512 bytes, then the executable E extracted from X is evaluated for its effects. By referring to the pre- and post-conditions of E, i.e., E.pre and E.post, we capture E's effects as if it were evaluated; this makes sense only for a resource that is an executable, and thus has an *eval* function defined for it. Note that most executables, when evaluated, will simply crash the MSHTML process.

After describing the vulnerability, we give a series of sample attacks, each of which shows how the vulnerability can be exploited by the adversary. Before giving some sample attacks for MS02-005a, we explain the parts in our template that we use to describe each attack.

Action Vulnerability: MSHTML processes HTML document D in zone Z.

Intended precondition: true
Actual precondition: D contains `<EMBED SRC=X>` $\Rightarrow$ length(X) $\leq$ 512
Intended postcondition: (one of many clauses)
D contains `<EMBED SRC=X>` $\wedge$ "Run ActiveX Controls " is enabled for Z $\Rightarrow$ display(X)
Actual postcondition: (one of many clauses)
D contains `<EMBED SRC=X>` $\wedge$ "Run ActiveX Controls" is enabled for Z $\Rightarrow$
[(length(X) $>$ 512 $\wedge$ extract_payload(X) = E) $\Rightarrow$ (E.pre $\Rightarrow$ E.post)
$\wedge$ length(X) $\leq$ 512 $\Rightarrow$ display(X)]

Attack 1: Web server executes arbitrary code on client.
Goal: Enable execution of arbitrary code on client.

Resource Table

Resource	*Carrier*	*Channel*	*Target/Enabler*
HTTPD web server (process)			E
server-client web connection C (data)		MP	E
browser B (process)			E
HTML document D (data)	Y		E
MSHTML (process)			T

Preconditions

- Victim requests a web page from adversary's site S.
- Victim's machine maps site S to zone Z.
- Victim's machine has "Run ActiveX Controls" security option enabled for zone Z.
- Adversary creates HTML document D containing an embed tag `<EMBED X>`, where length(X) $>$ 512 and extract_payload(X) = E.

Attack Sequence

1 Web server sends document D to browser B over connection C.

2 B passes D to MSHTML in zone Z.

3 **MSHTML processes D in zone Z.**

Postconditions

- Arbitrary, depending on the payload.

Figure 8.2. Microsoft Security Bulletin MS02-005a: Cumulative Patch for Internet Explorer (I)

Attack 2: Mail-based attack (HTML email) executing arbitrary code on client.
Goal: Enable execution of arbitrary code on client.

Resource Table

Resource	*Carrier*	*Channel*	*Target/Enabler*
HTTPD web server (process)			E
server-client mail connection C (data)		MP	E
Outlook Express OE (process)			E
HTML mail message M (data)			E
HTML document D (data)	Y		E
MSHTML (process)			T

Preconditions

- Victim able to receive mail from attacker.
- Victim's HTML email is received in zone Z.
- Victim's machine has "Run ActiveX Controls " security option enabled for zone Z.
- Adversary creates HTML document D containing an embed tag `<EMBED X>`, where length(X) > 512 and extract_payload(X) = E.
- Adversary creates mail message M with D included, where $Z \neq$ Restricted Zone.

Attack Sequence

1 Adversary sends HTML message M to victim via email.

2 Victim views (or previews) M in OE.

3 OE passes D to MSHTML in zone Z .

4 **MSHTML processes D in zone Z.**

Postconditions

- Arbitrary, depending on the payload.

Figure 8.3. Microsoft Security Bulletin MS02-005a: Cumulative Patch for Internet Explorer (II)

- The goal of the attack.
- A resource table showing for each resource (data or process) involved in the attack whether it serves as a carrier ("Y" means "yes; a blank, "no"), a channel (if so, "MP" means message-passing ; "SM" means shared-memory; and a blank means it is not a channel), or a target or enabler ("T" means it is a target; "E", an enabler).
- The pre-condition for the attack. Each clause is a conjunct of the pre-condition.
- The attack itself, written as a sequence of actions. The action exploiting the vulnerability is in boldface. (More formally, we would specify each action with pre- and post-conditions. For the attack to make sense, the pre-condition of the attack should imply the pre-condition of the first action in the attack, the post-condition of the ith action should imply the pre-condition of the $i + 1$st action, and the post-condition of the last action should imply the post-condition of the attack.)
- The post-condition for the attack. This post-condition corresponds to the adversary's goal, i.e., the reason for launching the attack in the first place. It should imply the goal (see first item above).

Let's now return to our example. Since MSHTML is used by both the browser and the mailer, we give two sample attacks, each exploiting the same vulnerability just described.

In the first attack (Figure 8.2), the adversary's goal is to run arbitrary code on the client. As indicated by the resource table for Attack 1, he accomplishes his goal by using the web server and the client browser as enablers. The server-client web connection is the message-passing channel by which the attack occurs. The HTML document is the carrier of the payload and the MSHTML process is the target of attack.

The pre-condition for the attack is that the victim should have requested a web page from the adversary and should have enabled for zone Z the option to run ActiveX Controls, and that the adversary's site is mapped to zone Z on the victim's machine. The attack itself is the sequence of three actions: the web server sends an HTML document D with an ill-formed embedded object to the client browser; the browser passes D to the MSHTML process; the MSHTML processes D as specified in the vulnerability. The post-condition of the attack is the effect of running the embedded executable.

In the second attack (Figure 8.3), the adversary's goal is the same and the vulnerability is the same. The means of attack, however, are different. Here, the enablers are an HTML mail document and the mailer process, i.e., Outlook Express. Note that people usually consider Outlook Express to be the target, but in fact, for this attack, it is an enabler. The channel, carrier, and target are the same as for the first attack.

The pre-condition is different: the victim needs to be able to receive mail from the attacker and HTML email received is in zone Z that is not the restricted zone. The attack is a sequence of four actions: the web server sends an HTML document D with an ill-formed embedded object to the victim via email; the victim views the HTML document in the mailer process, i.e., Outlook Express; the mailer process sends D to MSHTML in zone Z; and finally, the MSHTML processes D as specified in the vulnerability. The post-condition is as for the first attack, i.e., the effect of running the embedded executable.

5. Analyzing Attack Surfaces

We use our broad dimensions of targets and enablers, communication channels and protocols, and access rights to guide us in deciding (1) what things to count, to determine a system's attackability ; (2) what things to eliminate or reduce, to improve system security; and (3) how to compare two versions of the same system. In this section we consider briefly the first two items; Section 6 gives a detailed concrete example of all three.

5.1 Measuring the Attack Surface

We can define a measure of the system's attack surface to be some function of the targets and enablers, the channels associated with each type or instance of a target and enabler, the protocols that constrain the use of channels, and the access rights that constrain the access to all resources.

$$surf = f(targets, enablers, channels, protocols, access\ rights)$$

In general, we can define the function f in terms of additional functions on targets, enablers, channels, and access rights to represent relationships between these (e.g., the constraints imposed by protocols on channels, and the constraints imposed by access rights on all resources), or weights of each type (e.g., to reflect that certain types of targets are more critical than others or to reflect that certain instances of channels are less critical than others).

We deliberately leave f uninterpreted because in practice what a security analyst may want to measure may differ from system to system. Moreover, defining a precise f in general, even for a given system, can be extremely difficult. We leave the investigation of what different types of metrics are appropriate for f for future work. In Section 6 we give a very simplistic f.

5.2 Reducing the Attack Surface

The concepts underlying our attack surface also give us a systematic way to think about how to reduce it. We can eliminate or reduce the number of

(1) types or instances of targets, processes, enablers, executables, carriers, eval functions, channels, protocols, and rights; (2) types or instances of vulnerabilities, e.g., by strengthening the actual pre- or post-condition to match the intended; or (3) types or instances of attacks, e.g., through deploying one or more security technologies.

Principles and rules of thumb that system administrators and software developers follow in making their systems more secure correspond naturally to our concepts. For example, the tasks specified in "lockdown instructions" for improving security of a system frequently include eliminating data and process targets and strengthening access rights. Consider these examples:

Colloquial	Formal
Turn off macros.	Eliminate an eval function for one type of data.
Block attachments in Outlook.	Avoid giving any executable (data) as an argument to an eval function.
Secure by default.	Eliminate entire types of targets, enablers, and channels; restrict access rights.
Check for buffer overrun .	Strengthen the post-condition of the actual behavior to match that of the intended behavior.
Validate your input.	Strengthen the pre-condition of the actual behavior to match that of the intended behavior.
Change your password every 90 days.	Increase the likelihood that the authentication mechanism's pre-condition is satisfied.

6. An Example Attack Surface Metric

Howard identified a set of 17 RASQ vectors [Howard, 2003] and defined a simple attack surface function to determine the relative attack surface of seven different versions of Windows. In Section 6.1 we present 20 attack vectors: Howard's original 17 plus 3 others we added later. In Section 6.2 we present his RASQ calculation for all 20 attack vectors in detail. In Section 6.3 we analyze his RASQ results: we confirm observed behavior reflecting user experience and lockdown scenarios, but also we point out additional missing elements.

6.1 Attack Vectors for Windows

Howard's original 17 RASQ vectors [Howard, 2003] are shown as the first 17 in Figure 8.4. Upon our[5] initial analysis of his work, we noted that he had not considered enablers, such as scripting engines. Thus, we subsequently added three more attack vectors, shown in italics. Figure 8.4 shows how we map the 20 attack vectors into our terminology of channels, process targets, data targets, process enablers, and access rights.

We describe each in more detail below.

1 Open sockets: TCP or UDP sockets on which at least one service is listening. Since one service can listen on multiple sockets and multiple services can listen on the same socket, this attack vector is a channel type; the number of channels is independent of the number of services.

2 Open RPC endpoints: Remotely-accessible handlers registered for remote procedure calls with the "endpoint manager." Again, a given service can register multiple handlers for different RPC interfaces.

3 Open named pipes: Remotely-accessible named pipes on which at least one service is listening.

4 Services: Services installed, but not disabled, on the machine. (These are equivalent to daemons on UNIX systems.)

5 Services running by default: Services actually running at the time the measurements are taken. Since our measurements are taken when the system first comes up, these are the services that are running by default at start-up time.

6 Services running as SYSTEM: Services configured to log on as Local-System (or System), as opposed to LocalService or some other user. (LocalSystem is in the administrators group.)

7 Active Web handlers: Web server components handling different protocols that are installed but not disabled (e.g., the W3C component handles http; the nntp component handles nntp).

8 Active ISAPI filters : Web server add-in components that filter particular kinds of requests. ISAPI stands for Internet Services Application Programming Interface; it enables developers to extend the functionality provided by a web server. An ISAPI filter is a dynamic link library (.dll) that uses ISAPI to respond to events that occur on the server.

9 Dynamic web pages: Files under the web server root other than static (.html) pages. Examples include .exe files, .asp (Active Server Pages) files, and .pl (Perl script) files.

10 Executable vdirs: "Virtual Directories" defined under the web server root that allow execution of scripts or executables stored in them.

11 Enabled accounts: Accounts defined in local users, excluding any disabled accounts.

12 Enabled accounts in admin group: Accounts in the administators group, excluding any disabled accounts.

13 Null sessions to pipes and shares: Whether pipes or "shares" (directories that can be shared by remote users) allow anonymous remote connections.

14 Guest account enabled: Whether there exists a special "guest" account and it is enabled.

15 Weak ACLs in FS: Files or directories that allow "full control" to everybody. "Full control" is the moral equivalent of UNIX rwxrwxrwx permissions.

16 Weak ACLs in Registry: Registry keys that allow "full control" to everybody.

17 Weak ACLS on shares: Directories that can be shared by remote users that allow "full control" to everybody. Even if one has not explicitly created any shares, there is a "default share" created for each drive; it should be protected so that others cannot get to it.

18 VBScript enabled: Whether applications, such as Internet Explorer and Outlook Express, are enabled to execute Visual Basic Script.

19 Jscript enabled: As for (18), except for Jscript.

20 ActiveX enabled: As for (18), except for ActiveX Controls.

6.2 Attack Surface Calculation

In Howard's calculation, the attack surface area is the sum of independent contributions from a set of channels types, a set of process target types, a set of data target types, a set of process enablers, all subject to the constraints of the access rights relation, A.

$$surf^A = surf^A_{ch} + surf^A_{pt} + surf^A_{dt} + surf^A_{pe}$$

This simple approach has a major advantage in that it allows the categories to be measured independently. This simplification comes at a cost. For example, since interactions between services and channels are not considered, Howard's RASQ calculation fails to distinguish between sockets opened by a service running as administrator and (less attackable) sockets opened by a service running as an arbitrary user.

Figure 8.5 gives a table showing each of the four terms in detail. Each term takes the form of a double summation: for each type (of channel types, *chty*, process target types, *ptty*, data target types, *dtty*), and process enabler types, *pety*, for each instance of that type, a *weight*, ω, for that instance is added

20 RASQ Attack Vectors	Formal
Open sockets	channels
Open RPC endpoints	channels
Open named pipes	channels
Services	process targets
Services running by default	process targets, constrained by access rights
Services running as SYSTEM	process targets, constrained by access rights
Active Web handlers	process targets
Active ISAPI Filters	process targets
Dynamic Web pages	process targets
Executable vdirs	data targets
Enabled accounts	data targets
Enabled accounts in admin group	data targets, constrained by access rights
Null sessions to pipes and shares	channels
Guest account enabled	data targets, constrained by access rights
Weak ACLs in FS	data targets, constrained by access rights
Weak ACLs in Registry	data targets, constrained by access rights
Weak ACLs on shares	data targets, constrained by access rights
VBScript enabled	*process enabler*
Jscript enabled	*process enabler*
ActiveX enabled	*process enabler*

Figure 8.4. Mapping RASQ Attack Vectors into Our Formalism

to the total attack surface. For a given type, τ, we assume we can index the instances per type such that we can refer to the ith instance by τ_i. For *weight* functions, ω, that are conditional on the state of the instance (e.g., whether or not an account is default), we use the notation $(cond, v_1, v_2)$ where the value is v_1 if *cond* is true and v_2 if *cond* is false.

For channels, access control is factored into the weights in one very limited case: Howard gives a slightly lower weight to named pipes compared to the other channels because named pipes are not generally accessible over the Internet. An alternate, more general approach to modeling this situation would be to calculate a "local attack surface" and "remote attack surface," each of which is appropriate for different threats.

For process targets, the weight function for services makes use of the access rights relation explicitly by referring to whether a service is a default service or if it is running as administrator.

The influence of the access rights relation is the most obvious for data targets, since it is used to determine whether an account is in a group with administrator privileges and whether it is a guest account. Note that we view an account as a shorthand for a subset of the access rights, i.e., a particular principal with a particular set of rights. Access rights is also used to determine the value of *weakACL* on files, registry keys, and shares. The predicate *weakACL* is true of its data target if all principals have all possible rights to it, i.e.,"full control".

The weights for process enablers are the count of the number of applications that enable a particular form of attack. Here, we consider only two applications, Internet Explorer and Outlook Express; in general, we would count others. Script-based attacks, for example, may target arbitrary process or data targets, but are enabled by applications that process script embedded in HTML documents. Malicious ActiveX components can similarly have arbitrary targets, but any successful attack is enabled by an application that allows execution of the potentially malicious component.

Our reformulation of Howard's original model shows that there are only 13 types of attack targets, rather than 17; in addition, there are 3 types of enablers.

6.3 Analysis of Attack Surface Calculation

The results of applying these specific weight functions for five different versions of Windows are shown in Figure 8.1. As mentioned in the introduction, the two main conclusions to draw are that *with respect to the 20 RASQ attack vectors* (1) the default version of a running Windows Server 2003 system is more secure than the default version of a running Windows 2000 system, and (2) a running Windows Server 2003 with IIS installed is only slightly less secure than a running Windows Server 2003 without IIS installed.

$surf^A_{ch} = \sum_{c \in chty} \sum_{i=1}^{\|c\|} \omega(c_i)$	
chty	$\omega(c_i)$
socket	1.0
endpoint	0.9
namedpipe	0.8
nullsession	0.9

$surf^A_{pt} = \sum_{p \in ptty} \sum_{i=1}^{\|p\|} \omega(p_i)$	
ptty	$\omega(p_i)$
service	$0.4 + \mathit{def}(p_i) + \mathit{adm}(p_i)$
webhandler	1.0
isapi	1.0
dynpage	0.6

where $\mathit{def}(p_i) = (\mathit{default}(p_i), 0.8, 0.0)$
$\mathit{adm}(p_i) = (\mathit{run\_as\_admin}(p_i), 0.9, 0.0)$

$surf^A_{dt} = \sum_{d \in dtty} \sum_{i=1}^{\|d\|} \omega(d_i)$	
dtty	$\omega(d_i)$
account	$0.7 + \mathit{adg}(d_i) + \mathit{gue}(d_i)$
file	$(\mathit{weakACL}(d_i), 0.7, 0.0)$
regkey	$(\mathit{weakACL}(d_i), 0.4, 0.0)$
share	$(\mathit{weakACL}(d_i), 0.9, 0.0)$
vdir	$(\mathit{executable}(d_i), 1.0, 0.0)$

where $\mathit{adg}(d_i) = (d_i \in \text{AdminGroup}, 0.9, 0.0)$
$\mathit{gue}(d_i) = (d_i.\text{name} = \text{"guest"}, 0.9, 0.0)$

$surf^A_{pe} = \sum_{e \in pety} \sum_{e_i \in \{IE, OE\}} \omega(e_i)$	
pety	$\omega(e_i)$
vbscript	$(\mathit{app\_executes\_vbscript}(e_i), 1.0, 0.0)$
jscript	$(\mathit{app\_executes\_jscript}(e_i), 1.0, 0.0)$
activex	$(\mathit{app\_executes\_activex}(e_i), 1.0, 0.0)$

where IE = Internet Explorer
OE = Outlook Express

Figure 8.5. Howard's Relative Attack Surface Quotient Metric

While it is too early to draw any conclusions about Windows Server 2003, the RASQ numbers are consistent with observed behavior in several ways:

- Worms such as Code Red and Nimda spread through a variety of mechanisms. In particular, Windows NT 4.0 systems were at far greater risk of being successfully attacked by these worms if the systems were installed with IIS than if they were not. This observation is consistent with the increased RASQ of this less secure configuration.
- Windows 2000 security is generally perceived as being an improvement over Windows NT 4.0 security [IW, 2001]; the differences in RASQ for the two versions in a similar configuration (i.e., with IIS enabled) reflect this perception.
- Conversely, Windows 2000 (unlike Windows NT 4.0) is shipped with IIS enabled by default, which means that the default system is actually *more* likely to be attacked. This observation is consistent with anecdotal evidence that many Windows 2000 users (including one author of this paper) affected by Code Red and Nimda had no idea they were actually running IIS.

As a sanity check , we also measured the RASQ in two "lockdown" configurations: applying IIS security checklists to both NT 4.0 with IIS [MS-IISv4] and Windows 2000 [MS-IISv5]. Since the tasks specified in the lockdown instructions include disabling services, eliminating unnecessary accounts, and strengthening ACLs, the RASQ unsurprisingly decreases: on Windows NT 4.0, from 598.3 in the default configuration to 395.4 in the lockdown configuration; on Windows 2000, from 342.2 in the default to 305.1. These decreases are consistent with users' experience that systems in lockdown configurations are more secure; for example, such configurations were not affected by the Code Red worm [MSB, 2001].

Our set of 20 attack vectors still misses types and instances, some of which also need more complex weight functions:

- For channels, some IPC mechanisms were not counted; for example COM is counted if DCOM is enabled, but otherwise it is not.
- For process targets, we did not handle executables that are associated with file extensions that might execute automatically (i.e., "auto-exec") or be executed mistakenly by a user. Also, we did not count ActiveX controls themselves as process targets, only as process enablers, i.e., whether applications such as IE and OE were set up to invoke them.
- The model treats all instances of each type the same, whereas some instances should probably be weighted differently. For example, a socket

over which several complex protocols are transmitted should be a bigger contributor to the attack surface than a socket with a single protocol; and port 80 is well-known attack target that should get a higher weight than other channel endpoints.

- Just as for process targets that are services, for other types of process targets the weight function should take into consideration the privileges of the account that the process is executing as. For example, for versions of IIS ≤ 5.0, ISAPI filters always run as System, but in IIS 6.0, they run as Network Service by default.

These missing attack opportunities and refined weight functions suggest potential enhancements to Howard's RASQ model and the attack surface calculation.

7. Discussion of the RASQ Approach

We have some caveats in applying the RASQ approach naively:

- Obtaining numbers for individual attack vector classes is more meaningful than reading too much into an overall RASQ number. It is more precise to say that System A is more secure than System B because A has fewer services running by default rather than because A's RASQ is lower than B's. After all, summing terms with different units does not "type check". For example, if the number of instances in one attack vector class is N for System A and 0 for System B, but for a different attack vector class, the number is 0 for System A and N for System B, then all else being equal, the systems would have the same RASQ number. Clearly, the overall RASQ number does not reflect the security of either A or B with respect to the two different attack vector classes.

- The RASQ numbers we presented are computed for a given configuration of a running system. When an RASQ number is lower for System A than System B because certain features are turned off by default for System A and enabled by default for System B, that does not mean that System A is inherently "more secure"; for example, as the owner of System A begins to turn features on over time it can become just as insecure as System B. On the other hand, if 95% of deployed systems are always configured as System A initially (e.g., features off by default) and remain that way forever, then we could say in some global sense that we are "more secure" than if those systems were configured as System B.

- Do not compare apples to oranges. It is tempting to calculate an RASQ for Windows and one for Linux and then try to conclude one operating system is more secure or more attackable than the other. This would be a big mistake. For one, the set of attack vectors would be different for the two different systems. And even if the sets of attack vectors were identical, the threat models differ.

Rather, a better way to apply the RASQ approach for a given system is first to identify a set of attack vectors, and then for each attack vector class, compute a meaningful metric, e.g., number of running instances per class. Comparing different configurations of the same system per attack vector class can illuminate poor design decisions, e.g., too many sockets open initially or too many accounts with admin privileges. When faced with numbers that are too high or simply surprising, the system engineer can then revisit these design decisions.

8. Related Work

To our knowledge the notion of "attackability " as a security metric is novel. At the code level, many have focused on counting or analyzing bugs (e.g., [Chou et. al, 2001, Gray, 1990, Lee and Iyer, 1993, Sullivan and Chillarge, 1991]) but none with the explicit goal of correlating bug count with system vulnerability.

At the system level, Browne et al. [Browne et al., 2001] define an analytical model that reflects the rates at which incidents are reported to CERT. Follow-on work by Beattie et al. [Beattie et al., 2002] studies the timing of applying security patches for optimal uptime based on data collected from CVE entries. Both empirical studies focused on vulnerabilities with respect to their discovery, exploitation, and remediation over time, rather than on a single system's collective points of vulnerability.

Finally, numerous websites, such as Security Focus [SecurityFocus], and agencies, such as CERT [CERT] and MITRE [CVE], track system vulnerabilities. These provide simplistic counts, making no distinction between different types of vulnerabilities, e.g., those that are more likely to be exploited than others, or those relevant to one operating system over another. Our notion of attackability is based on separable types of vulnerabilities, allowing us to take relative measures of a system's security.

9. Future Work

Our state machine model is general enough to model the behavior an adversary attacking a system. We identified some useful abstract dimensions such as targets and enablers, but we suspect there are others that deserve consideration. In particular, if we were to represent *configurations* more explicitly, rather than as just states of the system (in particular the resources and access rights), then we can more succinctly define what it means for a process to be running by default or whether an account is enabled.

Further into the future we imagine a "dial" on the workstation display that allows developers to determine if they have just increased or decreased the attack surface of their code. We could flag design errors or design decisions that tradeoff performance for security. For example, consider a developer debating whether to open up several hundred sockets at boot-up time or to open sockets on demand upon request by authenticated users. For a long-running server, the first approach is appealing because it improves responsiveness and is a simpler design. However, even a simple attack surface calculation would reveal a significant increase in the server's attackability ; this potential security cost would need to be balanced against the benefits.

Measuring security, quantitatively or qualitatively, has been a long-standing challenge to the community. The need to do so has recently become more pressing. We view our work as a first step in revitalizing this research area. We suggest that the best way to begin is to start counting what is countable; then use the resulting numbers in a qualitative manner (e.g., doing relative comparisons). Perhaps over time our understanding will then lead to meaningful quantitative metrics.

Acknowledgments

This research was done while Jeannette Wing was a Visiting Researcher at Microsoft Research from September 2002-August 2003. She would like to thank Jim Larus, Amitabh Srivastava, Dan Ling, and Rick Rashid for hosting her visit.

This research is also now sponsored in part by the Defense Advanced Research Projects Agency and the Wright Laboratory, Aeronautical Systems Center, Air Force Materiel Command, USAF, F33615-93-1-1330, and Rome Laboratory, Air Force Materiel Command, USAF, under agreement number F30602-97-2-0031 and in part by the National Science Foundation under Grant No. CCR-9523972. The U.S. Government is authorized to reproduce and distribute reprints for Governmental purposes notwithstanding any copyright annotation thereon.

The views and conclusions contained herein are those of the authors and should not be interpreted as necessarily representing the official policies or endorsements, either expressed or implied, of the Defense Advanced Research Projects Agency Rome Laboratory or the U.S. Government.

Notes

1. For example, Microsoft's *Trustworthy Computing Initiative*, started in January 2002.
2. NT 4.0 measurements were taken on a system where Service Pack 6a had been installed; NT 4.0 with IIS enabled, with both Service Pack 6a and the NT 4.0 Option Pack installed. IIS stands for Internet Information Server.

3. There are more elegant formulations of composing two state machines; we use a simple-minded approach that basically merges two state machines into one big one. In the extreme, if the local resources sets are empty, then the two machines share all state resources; if the global resource set is empty, they share nothing. Thus our model is flexible enough to allow communication through only shared memory, only message passing, or a combination of the two.

4. Writing the return type of *eval* as *unit* is our way, borrowed from ML, to indicate that a function has a side effect.

5. Pincus and Wing

References

[Chou et. al, 2001] Andy Chou, Junfeng Yang, Benjamin Chelf, Seth Hallen, and Dawson Engler (2001). An empirical study of operating systems errors. In ACM Sym-posium on Operating Systems Principles, pages 73-88, October.

[Gray, 1990] J. Gray (1990). A census of tandem system availability between 1985 and 1990. IEEE Transactions on Software Engineering, 39(4), October.

[Lee and Iyer, 1993] I. Lee and R. Iyer (1993). Faults, symptoms, and software fault tolerance in the tandem GUARDIAN operating system. In Proceedings of the Inter-national Symposium on Fault-Tolerant Computing.

[Sullivan and Chillarge, 1991] M. Sullivan and R. Chillarge (1991). Software defects and their impact on system 118 availability. In Proceedings of the International Symposium on Fault-Tolerant Computing, June.

[SecurityFocus] Security Focus. http://www.securityfocus.com/vulns/stats.shtml.

[CERT] CERT. CERT/CC Advisories. http://www.cert.org/advisories/.

[CVE] MITRE. Common Vulnerabilities and Exposures. http://www.cve.mitre.org/.

[MS-IISv4] Microsoft TechNet (2001). Microsoft Internet Information Server 4.0 Security Checklist, July. http://www.microsoft.com/technet/security/tools /chklist/iischk.asp.

[MS-IISv5] Microsoft TechNet (2000). Secure Internet Informations Services 5 Checklist, June. http://www.microsoft.com/technet/security/tools/chklist /iis5chk.asp.

[MSB, 2001] Microsoft TechNet (2001). Microsoft Security Bulletin MS01-033, June. http://www.microsoft.com/technet/security/bulletin/MS-01-033.asp.

[Jampson, 1974] Butler Lampson (1974). Protection. Operating Systems Review, 8(1): pages 18-24, January.

[IW, 2001] Information Week (2001). Windows 2000 Security Represents a Quantum Leap, April. http://www.informationweek.com/834/winsec.htm.

[Howard, 2003] Michael Howard (2003). Fending OR Future Attacks by Reducing the Attack Surface, February. http://msdn.microsoft.com/library/default.asp?url=/library/en-us/dncode/html/secure02132003.asp.

[Lampson et al., 1992] Butler Lampson, Martin Abadi, Michael Burrows, and EdwardWobber (1992). Authentication in distributed systems: Theory and practice. ACM TOCS, 10(4):265-310, Novembe.

[MSRC] Microsoft Security Response Center. Security Bulletins. http://www.microsoft.com/technet/treeview/?url=/technet/security/current.asp?frame=true

[Schneider, 1991] Fred B. Schneider (1991). Trust in Cyberspace. National Academy Press, CSTB study edited by Schneider.

[Butler, 2003] Shawn Butler (2003). Security Attribute and Evaluation Method. PhD thesis, Carnegie Mellon University, Pittsburgh, PA.

[Beattie et al., 2002] Steve Beattie, Seth Arnold, Crispin Cowan, Perry Wagle, Chris Wright, and Adam Shostack (2002). Timing the application of security patches for optimal uptime. In 2002 LISA XVI, pages 101-110, November.

[Browne et al., 2001] Hilary Browne, John McHugh, William Arbaugh, and William Fithen (2001). A trend analysis of exploitations. In IEEE Symposium on Security and Privacy, May. CS-TR-4200, UMIACS-TR-2000-76.

[Pincus and Wing, 2003] Jon Pincus and Jeannette M. Wing (2003). A Template for Microsoft Security Bulletins in Terms of an Attack Surface Model. Technical report, Microsoft Research, in progress.

Chapter 9

A MODELING OF INTRUSION DETECTION SYSTEMS WITH IDENTIFICATION CAPABILITY

Pei-Te Chen
Department of Electrical Engineering
National Cheng Kung University
peder@crypto.ee.ncku.edu.tw

Benjamin Tseng
Department of Information Management
Hsing Kuo University of Management
btseng@mail.hku.edu.tw

Chi-Sung Laih
Department of Electrical Engineering
National Cheng Kung University
laihcs@eembox.ncku.edu.tw

Abstract In this paper, we first show that traditional IDSs cannot reach the minimal cost design from the auditing viewpoints. Then we propose the definition of design architecture of IDSIC (Intrusion Detection System with Identification Capability). In IDSIC, its architecture consists of a new detection engine that can examine packet headers, which provide a separability of security auditors and hackers. With this architecture, we will reduce the cost of false alarms, either positive or negative false alarms.

Keywords: Intrusion Detection System (IDS), Identification Capability, fingerprint , Security Auditor

1. Introduction

In 1987, Denning proposed the first model of intrusion-detection expert system (IDES) [Denning, 1987] that can detect a wide range of security, such as attempted break-in, masquerading or successful break-in, Trojan horse , virus , Denial-of-Service , and so on. After IDES model, the intrusion detection research topic has been a hot issue since the last decade. Several current systems are based on part of IDES prototype technology [Krawczyk et al., 1997, Bellare et al., 1996]. For example, the Next-Generation Intrusion Detection Expert System (NIDES) [Anderson et al., 1995], Multics Intrusion Detection and Alerting System (MIDAS) [Whitehurst, 1987], Network Anomaly Detection and Intrusion Reporter (NADIR) [Hochberg et al., 1993], Network Security Monitor (NSM) [Heberlein et al., 1990], Distributed Intrusion Detection System (DIDS) [Snapp et al., 1991], and so on.

However, most IDS models do not consider the role of security auditors. In some security standards, it suggests that there should be an inner auditor periodically checks the security issues in the enterprise networks. In order to discover the real security holes or vulnerabilities, the security tools using by the auditors are the same tools used by the outside hackers. These tests can be separated into two situations. One situation is the rehearsal; the auditors notify the system managers when the security auditing starts and how the security tests go on. It has to guarantee that no other attack can access the system during execution. Because both the system managers and the auditors know scenarios of security tests, the testing results in this situation are very little.

The other situation is that auditors imitate hackers' behaviors when performing security tests. The system managers do not know when, where, and how the tests will take place in advance. Because auditors use the same tools that used by hackers, the IDS response component will triggle alarms. If hackers attack the system in the meanwhile, system managers may ignore the real attack alarms. Thus, it not only takes system managers a lot of time to recover the system but also lowers the efficiency of IDSs during the security auditors examining the system.

Lee et al. propose a cost-sensitive model for IDSs by using some major cost factors, such as damage cost ($DCost$), response cost ($RCost$), operational cost ($OpCost$), etc, to evaluating the total cost of IDSs, and they mentioned that IDSs should minimize these costs [Fan et al., 2000, Lee et al., 2000]. However, the traditional IDSs (TIDSs) do not consider the behavior of the security auditors and the cost in TIDS is not the minimal. In this paper, we propose a new model of IDS, called IDSIC (Intrusion Detection System with Identification Capability), to distinguish the roles of hackers and auditors. Therefore, the damage cost and response cost in TIDS could be reduced. By this identification capability, IDSIC could minimize the associated cost mentioned in [Fan et al., 2000, Lee et al., 2000].

The rest of this paper is as follows. In Section 2, we describe the roles needed in traditional IDSs. The new model with identify capability will be proposed in Section 3. The key idea is from the IP packet marking technique

[Dean et al., 2001, Savage et al., 2000] using in IP traceback to detect the DDoS originator. Finally, a conclusion will be given in Session 4.

2. Traditional IDS model

2.1 Roles in TIDSs

In [Cannady, 2000], the design and implementations of TIDSs are focus on requirements, like tolerance, and minimal resource, etc. There are further requirements in large enterprise networks. One of which is the need of frequently audition testing. In order to describe these requirements in large networks, we give some definitions of the roles in TIDSs, so that we can discuss more general model in the next session. The roles in TIDSs could be classified into three parts.

Hackers - People who attempt to gain unauthorized access to a computer system. These people are often malicious and have many tools for breaking into a system.

System Manager (SM) - the person who takes charge to minimize the use o excess, network management, and system maintenance costs. If a system under some attacks results IDSs alarms, they have to make efforts to find out where the problem is.

Detection System (DS) - the system that monitor the events occurring i protected hosts or networks and analyze them for signs of intrusions. Figure 9.1 shows the relationships between these three roles.

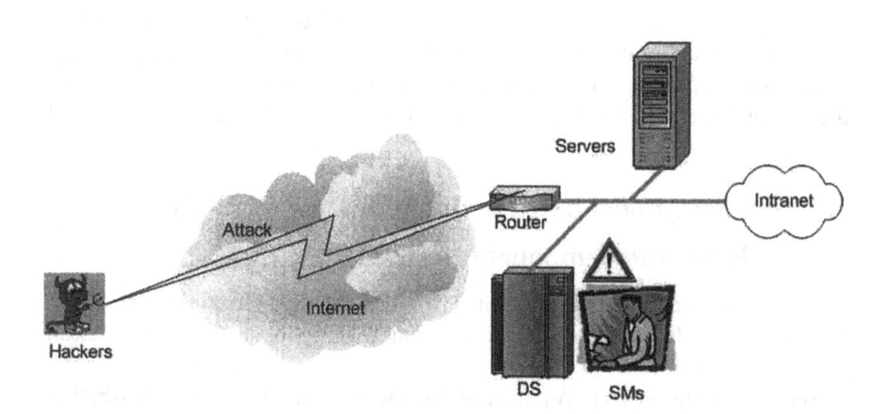

Figure 9.1. The roles and relationships in TIDSs.

2.2 Cost analysis

Table 9.1 shows the Consequential Cost (CCost) in every situation [Fan et al., 2000]. Here, event $e = (a, p, r)$ is the function of the attack a, the progress p, and the resource r. e' means the misidentified event. $PCost(e)$ represents penalty cost of treating legitimate event as an intrusion. ϵ_1 is the function of the progress p of the attack and $0 \leq \epsilon_1 \leq 1$. More detail analysis is shown in [Fan et al., 2000, Lee et al., 2000]. Therefore, for the entire event set E, and

Table 9.1. The Consequential Cost (CCost) in every situation [Fan et al., 2000, Lee et al., 2000]. Notice that FN, FP, TP and TN denote False Negative, False Positive, True Positive and True Negative respectively.

Situtation	Consequential Cost(CCost)	Condition
FN	$DCost(e)$	
FP	$RCose(e') + PCost(e)$ 0	$if DCost(e') \geq RCose(e')$ $if DCost(e') < RCose(e')$
TP	$RCose(e') + \epsilon_1 PCost(e), 0 \leq \epsilon_1 \leq 1$ $DCost(e)$	$if DCost(e) \geq RCose(e)$ $if DCost(e) < RCose(e)$
TN	0	

each event $e \in E$, we have the cumulative cost of TIDSs as follows:

$$CumulativeCost(E) = \sum_{e \in E} CCost(e) + OpCost(e)) \quad (1)$$

However, this cumulative cost is not the minimal cost in TIDSs. Because it does not think about the roles of the auditors performing security tests in large enterprise environment. By thinking of the auditing roles, the cost in (1) is not the minimal situation and can be reduced more. We will discuss this in the next Session.

3. A New model based on Identification (IDSIC)

3.1 Roles and components in IDSIC

We define a new Detection System with Identification Capability (DSIC) to tell the auditors from hackers. Roles and functionality of security auditors can be defined as follows.

Security Auditor (SA) - A person appointed and authorized to audit whether the security equipments work regularly or not by using the vulnerability testing tools.

Detection System with Identification Capability (DSIC) - One type of DS that runs the same function of DS. However, it has an extra functionality to distinguish between the roles of hackers and SAs.

When SAs want to perform the security tests, some secret information, called fingerprint, should be calculated in advance. We set two components, fingerprint adder and fingerprint checker, in SA and DSIC respectively. The fingerprint adder adds the fingerprint into the packets that are sent by the SAs. The relative component, fingerprint checker, helps DSIC to differentiate hackers' attack and SAs' tests. Figure 9.2 shows these roles and components in IDSIC.

3.2 Cost analysis

Since we take the role of SAs into account, all the cost in TIDS should be reconsidered. The damage cost ($DCost$) should be divided into two parts; hackers' and SAs' damage cost, i.e. $DCost(e) = HDCost(e) + SDCost(e)$

The term $HDCost(e)$ means the damage cost caused by hackers that may harm to the systems. The cost of SAs, $SDCost(e)$, is the amount of security testing cost that may damage to the systems. Similarly, the response cost ($RCost$) will also be separated into two parts: the cost of response generated by hackers ($HRCost$) and the one created by SAs ($SRCost$). It means that $RCost(e) = HRCost(e) + SRCost(e)$.

FN_{IC} Cost in IDSIC, this cost may be major in $HDCost$ although some serious security tests performed by SAs' will also harm to the systems. Therefore, the FNIC Cost can be presented by $HDCost(e) + \epsilon_3 SDCost(e)$. Here $\epsilon_3 \in [0, 1]$, and very close to 0, is a function of the progress p. FP_{IC} Cost happens when normal behavior, e, is misidentified as the attack, e' . By finger-

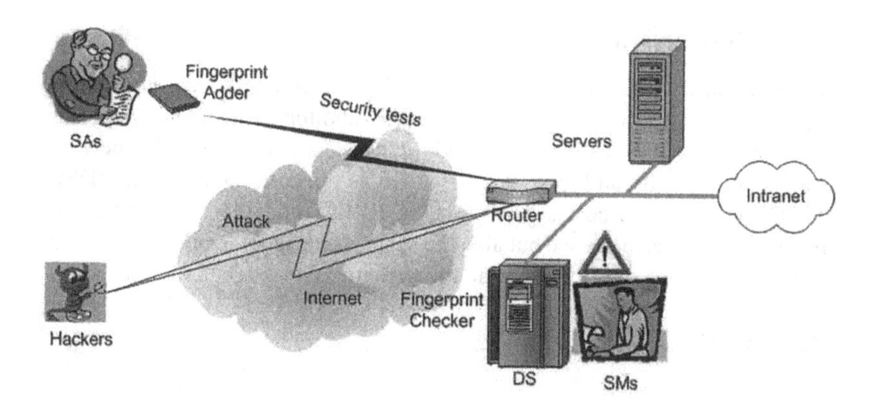

Figure 9.2. The roles and components in IDSIC.

print, we can distinguish SAs' events, e_{SA} , from these misidentified events, i.e. the remainder events, $e'' = e' - e_{SA}$, are real misidentified events. Hence, we could reduce the FP Cost by changing the misidentified event e' to e'' .

TP_{IC} Cost in IDSIC should compare with $HDCost$, $SDCost$, $HRCost$, and $SRCost$. Because DSIC does not alert alarms and just logs the occurrence when DSIC detects fingerprint sent by SAs, the $SRCost$ could be ignored. Besides, most security tests are not harmful to the system resource; the $SDCost$ could be near to 0. Therefore, the $DCost$ and $RCost$ in TIDS can reduce to $(HDCost(e) + \epsilon_4 SDCost(e)$ and $HRCost(e)$ respectively.

$$TP_{IC}Cost = \begin{cases} HRCost(e) + \epsilon_1 HDCost(e) + \epsilon_1\epsilon_4 SDCCost(e), \; 0 \le \epsilon_1, \epsilon_4 \le 1 & if(HDCost(e) + \epsilon_4 SDCost(e)) \ge HRCost(e) \\ HDCost(e) + \epsilon_4 SDCost(e) & if(HDCost(e) + \epsilon_4 SDCost(e)) < HRCost(e) \end{cases}$$

Same with TIDSs, TN Cost should be always 0 that DSIC detects no attacks. Therefore, we could minimize the IDSIC's CumlativeCost by rewriting the $CCost(e)$ to $ICCost(e)$ in Equation (1). The following Equation (2) shows this result. The ICCost(e) is defined in Table 2.

$$CumulativeCost(E) = \sum_{e \in E} ICCost(e) + OpCost(e)) \quad (2)$$

Comparing with Equation (1) and (2), although $OpCost(e)$ is equivalent, but $CCost(e)$ in TIDS is greater than $ICCost(e)$ in IDSIC in every situation. Therefore, by fingerprint, IDSIC could have smaller CumulativeCost(E) than TIDS.

4. Conclusion

In this paper, we propose that there should exist a role, security auditor, inspecting the system holes. From the security auditors' viewpoint, we present a new model of IDS, called IDSIC, which can distinguish the roles between hackers and security auditors. We also analyze the costs in TIDS and IDSIC and IDSIC has lower costs. Therefore, in IDSIC model, we can ensure not only the system performance but also the system managers' working efficiency during the security auditors examine the system.

References

[Anderson et al., 1995] D. Anderson, T. Frivold, and A. Valdes (1995). Next-generation Intrusion Detection Expert System (NIDES) A Summary.

Technical Report SRI-CSL-95-07, SRI Computer Science Laboratory, May.

[Denning, 1987] D. E. Denning (1987), An intrusion-detection model. IEEE Trans. on Software Enginering, vol. SE-13, no.2, pages 222V232, February.

[Dean et al., 2001] D. Dean, M. Franklin, and A. Stubblefield (2001). An algebraic approach to ip traceback. In Network and Distributed System Security Symposium, NDSS '01, February.

[Krawczyk et al., 1997] H. Krawczyk, M. Bellare, and R. Canetti (1997). HMAC: Keyed-Hashing for Message Authentication. Internet RFC 2104, February.

[Cannady, 2000] J. Cannady (2000). An Adaptive Neural Network Approach to Intrusion Detection and Response. Ph.D Thesis, Nova Southeastern University.

[Hochberg et al., 1993] J. Hochberg, K. Jackson, and C. Stallings et al. Nadir (1993): An automated system for detecting network intrusion and misuse. In Computers and Security, vol.12, no.3, pages 235-248, May.

[Heberlein et al., 1990] L. Heberlein, G. Dias, and K. Levitt (1990). A Network Security Monitor. In Proceedings of the IEEE Symposium on Research in Security and Privacy, pages 296-304, May.

[Bellare et al., 1996] M. Bellare, R. Canetti, and H. Krawczyk (1996). Message authentication using hash functions: The HMAC construction. In RSA Laboratories' CryptoBytes Vol. 2, No. 1, Spring.

[Whitehurst, 1987] R. A. Whitehurst (1987). Expert systems in intrusion-detection: A case study. Technical report, Computer Science Laboratory, SRI International, Menlo Park, California, November.

[Savage et al., 2000] S. Savage, D.Wetherall, A. Karlin, and T. Anderson (2000). Practical network support for IP traceback. In 2000 ACM SIGCOMM Conference, August.

[Snapp et al., 1991] S. Snapp, J. Brentano, and G. Dias (1991). DIDS (Distributed Intrusion Detection System) – Motivation, Architecture, and An Early Prototype. In Proc., 14th National Computer Security Conference, Washington, DC, pages 167-176, October.

[Fan et al., 2000] W. Fan, W. Lee, S. Stolfo, and M. Miller (2000). A multiple model cost-sensitive approach for intrusion detection. In Proc. Eleventh European Conference of Machine Learning, pages 148-156, Barcelona Spain, May.

[Lee et al., 2000] W. Lee, M. Miller, and S. Stolfo (2000). Toward cost-sensitive modeling for intrusion detection. Technical Report CUCS-002-00, Computer Science.

Chapter 10

A SOURCE-END DEFENSE SYSTEM AGAINST DDOS ATTACKS

Fu-Yuan Lee
Department of Computer Science and Information Engineering, National Chiao Tung University, Hsinchu, Taiwan 300

Shiuhpyng Shieh
Department of Computer Science and Information Engineering, National Chiao Tung University, Hsinchu, Taiwan 300
URL: http://dsns.csie.nctu.edu.tw/ssp
E-mail:ssp@csie.nctu.edu.tw

Jui-Ting Shieh
Department of Computer Science and Information Engineering, National Chiao Tung University, Hsinchu, Taiwan 300

Sheng-Hsuan Wang
Department of Computer Science and Information Engineering, National Chiao Tung University, Hsinchu, Taiwan 300

Abstract In this paper, a DDoS defense scheme is proposed to deploy in routers serving as the default gateways of sub-networks. Each router is configured with the set of IP addresses belonging to monitored sub-networks. By monitoring two-way connections between the policed set of IP addresses and the rest of the Internet, our approach can effectively identify malicious network flows constituting DDoS attacks, and consequently restrict attack traffics with rate-limiting techniques. Current source-end DDoS defense scheme cannot accurately distinguish between network congestion caused by a DDoS attack and that caused by regular events. Under some circumstances, both false positive and false negative can be high, and this reduces the effectiveness of the defense mechanism. To improve the effectiveness, new DDoS detection algorithms are presented in this paper to complement, rather than replace existing source-end DDoS defense systems. The design of the proposed detection algorithm is based on three essential characteristics of DDoS attacks: *distribution*, *congestion*, and *continuity*. With the three characteristics, the proposed detection algorithm significantly improves detection accuracy, and at the same time reduces both false positive and false negative against DDoS attacks.

Keywords: information warfare , DoS/DDoS attacks , source-end defense

1. Introduction

Current Internet infrastructure is vulnerable to network attacks, and particularly, many security incidents have shown that the Internet is weak against distributed denial-of-service (DDoS) attacks. In general, a DDoS attack is accomplished by persistently overloading critical resources of the target Internet service so as to completely disable or degrade the service over an extended period of time. Such resource overloading can be achieved in several ways. First, an Internet service can be overloaded by a large number of service requests issued in a short period of time. As a result, legitimate service requests may be dropped due to insufficient resource such as computation power or memory space. Second, attackers can overload a network link near the target, and consequently, all flows traverse through the link will experience significant degrade of service quality .

To generate a great amount of traffic or service requests, attackers may first compromise a large number of computer systems. This can be easily accomplished due to the large number of insecure computer systems and the set of easily acquired and deployed exploit programs, such as Tribal Flood Network (TFN), TFN2K and Trinoo . On the other hand, detecting or preventing a DDoS attack is relatively much harder. The lack of explicit attack signatures/patterns makes it extremely difficult to distinguish attacks from legitimate traffic. Furthermore, the anonymous nature of IP protocol allows the attackers to disguise the attack origins, and thus makes it hard to detect the sources of DDoS attacks. These difficulties make the construction of an effective DDoS defense mechanism become a very challenging problem.

Issues for defending DDoS attacks have been extensively investigated in recent years, and several defense mechanisms have been presented in the literature. The deployment of these schemes can be categorized into three classes. The first class of schemes [Shaprio and Hardy, 2002, T. Aura and Leiwo, 2001, Mirkovic et al., 2002a, Juels and Brainard, 1999, Wang and Reiter, 2003, Leiwo et al., 2000, Mann et al., 2000, Feinstein et al., 2003, NFR, , Net, , Roesch, 1999] involve detecting and preventing a DDoS attack at the victim network. In this context, the term *victim network* indicates that the installed DDoS defense systems are used to protect a limited set of computers. These defense systems are generally deployed at end host systems or at routers which are able to examine and control communications between protected hosts/networks and the rest of the Internet. Placing defense mechanisms at the victim networks can be easier for detecting DDoS attacks. Since the DDoS traffic is aggregated toward the victim, a burst of network traffic would be the signal of a DDoS attack. However, from the network's perspective, filtering out DDoS attack packets at the victim side is ineffective because the attack flows may cause network congestion and waste valuable computation power of the routers along the path they traversed.

To improve the effectiveness of packet filtering, schemes in the second class attempt to construct DDoS defense lines toward attack sources. To achieve this objective, several schemes have been proposed, and these schemes can be further divided into two types. First, DDoS attacks are detected by DDoS defense systems installed in victim networks , and subsequently Internet core routers in the attack paths are requested to filter out attack traffic according to filtering criteria specified by downstream routers or DDoS defense systems [Ferguson, 1998, Park and Lee, 2001, Sung and X, 2002, Ioannidis and Bellovin, 2002, man, , Mahajan et al., 2002]. Second, traceback techniques [Savage et al., 2001, Savage et al., 2000, Dean et al., 2002, Song and Perrig, 2001] are utilized to identify attack sources and then legal sanctions can be performed to deter DDoS attacks. Schemes in the second class can partially avoid attack flows blending with legitimate flows and consequently somewhat reduce the complexity for distinguishing from attack traffic and legitimate traffic. Furthermore, it may also reduce to certain degree possible network congestion caused by attack flows. However, owing to the cooperative and distributed nature, these schemes heavily rely on cooperation among Internet core routers . This would generally incur high deployment cost. Routers need to be upgraded to support packet filtering in high speeds. Coordination among ISPs may also bring unpredictable difficulties. In addition to the deployment costs, the way that core routers drop packets according to the information passed from victim-end systems may implicitly bring other substantial cost and security breaches. For instance, an Internet-wide authentication framework is needed; otherwise, core routers may accept instructions from malicious attackers and drop legitimate traffic. Therefore, to secure and authenticate communications between core routers and victim-end systems in large networks may bring infeasible high overhead. Thus, schemes in the second class are generally inadequate to be deployed in large networks such as the Internet.

Similar to the victim-end approaches, the third class of schemes involve deploying DDoS defense mechanisms at default gateways. The major difference is that, DDoS defense mechanisms in the third class are used to police hosts in the monitored networks from participating in DDoS attacks rather than protecting them. This approach can ideally prevent attack traffic from entering the Internet. In other words, DDoS attack flows are contained in their sources.

It subsequently avoids attack flows blending with legitimate flows, and as a result network congestion can be significantly reduced. Furthermore, since the degree of flow aggregation is relatively low and routers closer to source networks are likely to relay less traffic than core routers, it is possible to use sophisticated detection strategies which may require more computation power and system resources.

Although the idea of defending DDoS at sources is attractive, detecting the occurrence of a DDoS attack at the attack sources is very difficult [Chang, 2002]. The main difficulty arises from the insignificant aggregate of attack traffic which can be observed in attack sources. Other criteria for identifying DDoS attacks must be discovered. For instance, in the D-WARD system proposed by *Mirkovic et al* [Mirkovic et al., 2002b], network congestion measured by the ratio of incoming and outgoing packets of network connections is used to judge whether the monitored flow is part of a DDoS attack or not. By monitoring the changes of the ratio, D-WARD would be able to detect a DDoS attack that has already disable the victim. However, it is hard for D-WARD to distinguish a DDoS attack from network congestion caused by other events. On one hand, D-WARD can mis-classified a flow if the ratio of flow is high in its normal operation. On the other hand, D-WARD is weak in detecting low-rate attacks . In other words, a well-designed attack script can avoid being detected by D-WARD by carefully control the congestion caused by the attack.

To address the weakness of D-WARD, in this paper, we propose a source-end DDoS detection algorithm and an attack response mechanism, where the former can accurately identify an ongoing DDoS attack and the latter can effectively limit attack traffic in source networks. The proposed detection and response algorithms are built upon the system architecture originally proposed in D-WARD. Our proposal focuses on reducing both false positive and false negative on detecting two-way connections. That is, the proposed scheme attempts to complement, rather than replace the D-WARD system.

The design of proposed scheme is based on the observation of three essential characteristics of a DDoS attack: *distribution*, *congestion*, and *continuity*. *Distribution* refers to the spreading of attack traffic from a large number of compromised hosts. *Congestion* refers to the inherent consequence of a DDoS attack. That is, an increasing packet loss rate observed in a monitored network flow would generally represent a signal of a DDoS attack . Third, *continuity* directs to the observation that network congestion caused by DDoS attacks usually lasts for an extended period of time. Combining the above three criteria allows us to differentiate a DDoS attack from a typical network congestion caused by other events. Based on the three characteristics, a new DDoS defense mechanism is proposed. Since the proposed mechanism is built upon D-WARD architecture, the proposed DDoS defense mechanism is also deployed at routers serving as the default gateways. Online traffic statistics, in terms of distribution, congestion, and continuity, are gathered and compared against previous statistics derived from normal traffic. In this way, malicious network flows are identified and rate-limited. Rate limits are dynamically adjusted according to the behavior of malicious network flows. On one hand, dynamic adjustment allows a misclassified network flow to regain network bandwidth when the flow shows compliance to legitimate flow model. On the other hand, since attack scripts has no way to distinguish the effect of rate-limiting from that of a successful DDoS attack, dynamic adjustment helps restrain malicious flows.

This paper is organized as follows. Section 2 gives an review of the D-WARD system. The proposed source-end DDoS defense scheme is presented in Section 3, including its detection and rate-limiting mechanism. Section 4 describes an implementation of the proposed scheme and presents several experiments on estimating the effectiveness. Subsequently, we summarize and conclude our findings in Section 5.

2. Review of D-WARD

In this section, we briefly review D-WARD system, including system architecture, detection algorithm, and attack response algorithm.

2.1 System Architecture

From the architectural point of view, D-WARD consists of a *observation component* and a *throttling component*. The observation component examines all communications between the set of IP addresses in the monitored network and the external IP addresses, and then computes on-line traffic statistics. Note that, in D-WARD, time are divided into a set of uniform intervals, called *observation period* , which serves as a unit time frame to compute traffic statistics. In each observation period, new traffic statistics are compared against past statistics derived from normal traffic. Network flows are classified according to the comparison results. Moreover, the statistics and comparison results are then passed to the throttling component which generates rate-limiting rules based on the behavior of the monitored network flows.

Fig. 10.1 shows a possible deployment of D-WARD. As depicted in the figure, D-WARD is a separate unit that acquires traffic from the default gateway and feeds the gateway with rate-limiting rules.

2.2 Attack Detection

In D-WARD, the aggregate traffic between monitored addresses and a correspondent host is defined as a *flow*. A flow is considered two-way if its data flow comprises packets originating from the sender and corresponding reply from the peer. TCP connections and several types of ICMP messages, such as

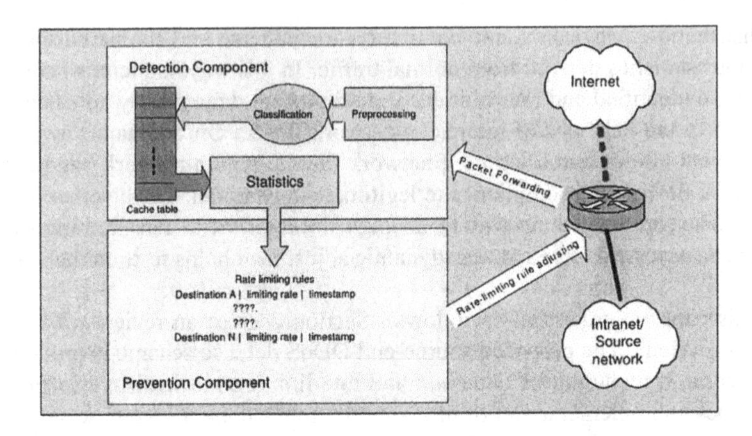

Figure 10.1. An example of the deployment of D-WARD

"timestamp" and "echo", are typical examples of two-way flows. On the other hand, a flow is considered one-way (or uni-directional) if its data flow does not require reply messages in its normal operation. Traffics based on UDP protocol are examples of one-way flows.

For TCP flows, D-WARD defines a threshold that specifies the maximum allowed ratio of the number of packet sent to and received from in a flow. Notice that in the following context in this paper, the ratio of the number of packet sent and received in a flow is refereed to as the *O/I* of the flow. Then, for TCP flows, whenever the O/I value of a flow breaches a pre-defined threshold, TCP_{rto}, the flow is classified as a DDoS attack flow. Similarly, for ICMP-based two-way flows, $ICMP_{rto}$ is used to define the maximum O/I value of an ICMP flow. In D-WARD's experimental settings, TCP_{rto} is set to 3 and $ICMP_{rto}$ is set to 1.1.

For a one-way flow, D-WARD defines three thresholds: an upper bound on the number of allowed monitored hosts issuing one-way connections to a correspondent host, a lower bound on the number of allowed packets in each one-way connection, and a maximum allowed sending rate. Whenever any of the three threshold is breached, the flow is considered attack. In D-WARD's experiment, the number of host in the same UDP flow must be smaller than 100. There must be at least one packet in each connection, and the maximum allowed sending rate is 10MBps.

In D-WARD, a flow is classified as *normal*, *suspicious* or *attack* according to the comparison on the statistics derived from normal flows and the currently gathered statistics. If the statistics of a monitored flow does not consistent with normal model defined by thresholds mentioned above, the flow is classified as attack. If a flow that was ever classified as attack and the current comparison indicates compliance with normal flow model, it is classified as suspicious. Finally, if a flow is always compliant with normal flow model, it is classified as normal.

2.3 Attack Response

According to the comparison result passed from the observation component, the throttling component specifies allowed sending rates for monitored flows. D-WARD utilizes a flow control mechanism which is similar to the congestion control mechanism of TCP protocol. The sending rate is exponentially decreased in the first phase of attack response. Then, if further comparison indicates compliance with the normal flow model, a rate-limited flow can regain its bandwidth after the slow recovery and fast recovery process. On the other hand, a rate-limited flow can be more severely restrained if it does not comply with the rate limit and attempts to persistently rebel against the limited sending rate.

3. Proposed System

In this section, we first show that there are normal TCP flows with its O/I value which is greater than the threshold defined by D-WARD. This indicates that D-WARD would classify these TCP flows as attack while they are in their normal operations. This problem cannot be solved by using a sufficient large threshold since it will increase the false negative. Specifically, low rate attacks will not be detected. To cope with the problem, a new algorithm for detecting and limiting TCP-based DDoS attacks are presented herein.

It is worthy to notice that although DDoS attacks may take many different forms, it is reported [Chang, 2002, Mahajan et al., 2002, Moore et al., 2001] that over 94% of DDoS attacks use TCP. Thus, the scheme presented in this paper may help defend against a majority of DDoS attacks. As to the detection of DDoS attacks based on of one-way flows, we suggest using the algorithm presented in D-WARD at current stage, but further enhancement is possible for the future work.

3.1 Basic idea of the proposed scheme

As mentioned above, D-WARD classifies a TCP flow as an attack flow if the O/I value of the flow is greater than TCP_{rto}. (Recall that, in D-WARD, this threshold is set to 3.) This approach suffers from the difficulty in determining an appropriate value for TCP_{rto}. It is because the O/I value of a TCP flow heavily depends on the implementation of TCP/IP protocol stack of the peers, and other factors such as round trip time and network congestion . This would result in a wide range of O/I values. For instance, Fig. 10.2 shows the average O/I values of TCP flows in a typical network consisting of 30 personal computers. Operating systems installed in these computers include Windows 2000, Windows XP, FreeBSD, and Linux. As shown in the figure, there are flows with O/I values which are greater than 3. (The highest average O/I value is 3.68. It is observed in a flow consisting of only one FTP data connection.)

The observation motivates a new algorithm for detecting TCP-based DDoS attacks. The proposed algorithm exploits three essential characteristics of DDoS attacks, namely distribution, congestion and continuity, to detect the presence of DDoS attacks. First, distribution refers to the observation that DDoS attack scripts will normally infect as many insecure computer systems as possible so as to amplify the power of the DDoS attack. Therefore, in the monitored networks, if there is an increasing number of hosts attempting to send traffic to a destination host, a DDoS attack may just have been started. The statistics on the number of hosts sending packets to the same target will provide a valuable criterion for judging whether there is a DDoS attack or not. Second, DDoS attack usually lead to high packet loss rate toward the victim. Since monitoring packet loss rates of individual TCP flows would incur infeasible high cost, similar to D-WARD, the packet loss rate of a flow is abstractly represented as the O/I value of the flow. Third, continuity reflects to the observation that a DDoS attack usually lasts for an extended period of time. As we shall see later, this makes us be able to distinguish network congestion caused by DDoS attacks and other network events.

By taking advantage of the three DDoS characteristics, the proposed detection algorithm can classify TCP flows more precisely. In the proposed scheme, there are two phases: initialization phase and detection phase. In the initialization phase, the proposed scheme constructs initial profiles for TCP flows according to the past traffic in the flows. In the profile database, each profile specifically represents the legitimate flow model of a TCP flow. Then, in the detection phase, traffic statistics are then compared with profiles. Profiles are

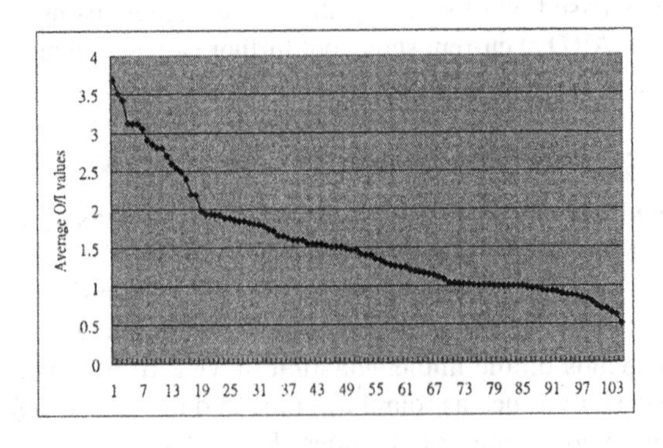

Figure 10.2. Average O/I values

dynamically adjusted to reflect the current behavior of monitored flows In this way, different thresholds can be used to classify different TCP flows, and thus the efficiency of the detection algorithm can be effectively improved.

3.2 Detection Phase

In the proposed scheme, characteristics and thresholds of a flow are derived from the past traffic of the monitored flow. For each TCP flow, its traffic statistics computed from the current observation period are compared against the legitimate flow model defined by the profile of the flow so as to determine whether it is malicious or not. To better understand the proposed legitimate flow model, some notations are introduced as follows.

First, a two-way flow f is a collection of connections, and each connection is associated with a pair of IP addresses – an IP address in the set of monitored addresses and an IP address of the correspondent hosts. The former is referred to as *initial address* and the latter is *terminal addresses*. The number of distinct initial addresses in a flow f is denoted as S_f. For a connection c, n_c denotes the ratio of the number of packets originated from the initial address and received from the terminal address in one observation period in connection c. Then, n_f represents the average of the O/I value of all connections in flow f.

Furthermore, there are two threshold values, N_f and T_f, which help determine the malicious level of a monitored flow. N_f represents the mini threshold of a flow f. If $n_f \leq N_f$, then f is considered as a normal flow. T_f denotes the maximum allowed n_f. If $n_f \geq T_f$, then f is classified as an attack flow. If $N_f \leq n_f \leq T_f$, then further traffic statistics must be examined to determine the malicious level of the flow.

Then, the level of congestion and distribution can be quantified. Consider a flow f with $N_f \leq n_f \leq T_f$, the level of congestion of f refers to $(n_f - N_f)/(T_f - N_f)$. In this expression, we can clearly see that if the packet loss rate of the flow approaches T_f, the value of the expression will approach 1. On the other hand, if n_f approaches N_f, the value will approach 0. Next, the level of distribution is quantified as S_f/C, where C denotes a configuration parameter obtained from the past behavior of the monitored network (We will describe how to obtain this parameter later). Then, the level of congestion and distribution are combined and used to generate a value representing the malicious level of a monitored flow. Herein, the malicious level is denoted α and computed as follows. (In Eq. 10.1, λ is a number between 0 and 1, that is, $0 < \lambda < 1$. It is used to restrict the saturation of α between 0 and 1.)

$$\alpha = \frac{1-\lambda}{\lambda} * \sum_{i=1}^{\lfloor S_f/C \rfloor} (\lambda * \frac{n_f - N_f}{T_f - N_f})^i \qquad (10.1)$$

It is worthy to note that α has two important characteristics. First, it is clear that α increases as n_f increases. In other words, if the packet loss rate of a monitored flow f gets higher, n_f will increase and consequently α increases. Second, α increases along with S_f even if n_f remains the same. This feature is especially useful in detecting DDoS attacks launched by attack programs which spoof source IP addresses. The α value will close to 0 when the monitored flow is in its normal operation. On the other hand, it will increase significantly when both the level of congestion and the level of distribution increases.

Although a surge of α value may indicate an DDoS attack that results in an abnormal increase in the packet loss rate or in the number of initial addresses in a flow, the α value can also go up due to regular network congestion. Nevertheless, the period of time the α value arises becomes a significant difference between the two causes. That is, normal network applications will stop sending more packets to a highly congested destination host after several attempts while DDoS attack scripts continually flush the victim for an extended period of time. With this observation, we can effectively distinguish a DDoS attack from a conventional network congestion by examining the length of time that DDoS attack signal lasts. This concept is implemented as follows. Consider a TCP flow f. α_f is a threshold that represents the maximum allowed α derived from the current network traffic. Once the threshold α_f is breached consecutively for t_f observation periods, f is considered a DDoS attack flow.

According to the proposed DDoS detection strategy, a network flow f can be classified into four types: *normal*, *suspicious*, *attack*, and *transient*. The transition of these types are depicted in Fig. 10.3. In brief, f is classified as a suspicious flow if $\alpha \geq \alpha_f$, where α is derived from the traffic in the current observation period. If α_f is breached for consecutive t_f observation periods, f is classified as an attack flow, and rate limiting techniques are applied to f. Once the traffic statistics of f shows compliance with legitimate flow model, i.e. $\alpha \leq \alpha_f$, for consecutive *PenaltyPeriod* observation periods, f is then classified as transient. For transient flows, rate limiting rules are carefully removed. When the allowed bandwidth of f reaches *MaxRate*, f is classified as a normal flow. Algorithm 1 shows pseudo code of the proposed detection algorithm.

In addition to the determination of the malicious level of monitored flows, it is desirable to update the thresholds for classifying network flows. This allows our scheme to learn the changing behavior of normal traffic, and dynamically adjust the thresholds according the current traffic statistics of monitored flows. For the adjustment of thresholds, attack traffic will be filtered out, and only traffic of a normal flow will be used to update thresholds. In this way, thresholds will not be polluted by attack traffic.

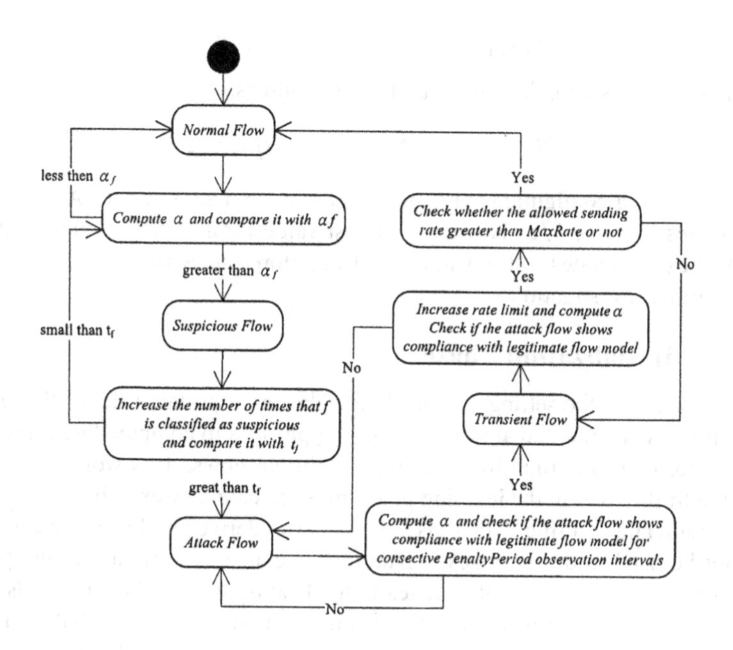

Figure 10.3. Classification of Traffic Flow

Updating a threshold is accomplished by feeding back the current traffic statistics. As we shall see shortly in this section, given a volume of historical traffic of the monitored flow f, we can derive T_f, and N_f for the monitored flow. In fact, by the same procedure, we can compute traffic statistics for each observation period. Assume that flow f is classified as normal in period i. Let $\widehat{T}_{f,i}$ denote the maximum allowed O/I value derived from the traffic volume in observation period i, and $T_{f,i}$ denote the same threshold used to classify flow f in period i (i.e., T_f of period i). Then, $T_{f,i+1}$ of the next observation period $(i+1)$ can be computed in the same way:

$$T_{f,i+1} = \beta * \widehat{T}_{f,i} + (1 - \beta) * T_{f,i}$$

In the similar fashion, N_f can be updated as follows.

$$N_{f,i+1} = \beta * \widehat{N}_{f,i} + (1 - \beta) * N_{f,i}$$

where β is a configurable parameter ranging from zero to one. At present, we suggest β to be $1/2$. However, its best value for a particular type of networks heavily relates to the variation of monitored network traffic, and may need further investigation.

3.3 Initialization Phase

It is clear that the settings of thresholds play an important role for the classification. As mentioned above, the learning phase is to compute these thresholds according to normal traffic of the monitored flows. It is worthy to note that the traffic used in the learning phase must be carefully examined and cannot contain attack traffic. Otherwise, the statistics derived will be incorrect and cannot be used to detect attacks. Due to this concern, Currently in the proposed scheme we perform off-line learning. That is, after all the thresholds are determined according to the historical traffic, the learning phase halts. This helps prevent our scheme from being polluted by on-line attack traffic. Next, the configurations of the threshold are described.

Consider parameters used in Eq. 10.1, i.e. T_f, N_f and C. Recall that T_f stands for the maximum allowed threshold of the O/I value of a monitored flow and N_f represents a mini threshold of the O/I value. Assume that the trail of historical network traffic is available which does not contain attacks. The traffic is partitioned into volumes in terms of observation periods, and then, thresholds are derived from the partitioned traffic volumes. In our scheme, the O/I value of each observation is measured. Then, we set $T_f = 2 * OI_{f,max}$ and $N_f = OI_{f,avg}$, where $OI_{f,max}$ denotes the maximum observed O/I value of flow f derived from historical traffic data and $OI_{f,avg}$ denotes the average O/I value. Next, Let C be the maximum number of distinct initial addresses in a flow during one observation period.

Algorithm 1 Detection Procedure

```
1: loop
2:   Collect IP packets received in one observation period.
3:   for each packet originating from monitored IP addresses do
4:     if Protocol = TCP then
5:       Classify the packet into a flow according to the destination IP address.
6:     end if
7:   end for
8:   calculate the O/I values of monitored TCP connections.
9:   for each flow (let the current flow be denoted as f) do
10:    if N_f ≤ n_f ≤ T_f then
11:      compute α for flow f.
12:      if α ≥ α_f then
13:        increase the number of time that f is classified as suspicious.
14:        if the number of times that f is classified as suspicious ≥ t_f then
15:          generate a DDoS attack alert and classify the flow as attack.
16:          perform rate-limiting.
17:        end if
18:      else
19:        reset the number of times that f is classified as suspicious.
20:      end if
21:    else if n_f ≥ T_f then
22:      Set α to 1, generate a DDoS attack alert and classify the flow as attack.
23:      perform rate-limiting
24:    end if
25:  end for
26: end loop
```

After T_f, N_f and C are configured, we can then compute a set of α values, one for each observation period. Then, we can set α_f to the average of the set of α values and subsequently set t_f to be the maximum consecutive number of times that α_f is breached in the set.

3.4 Rate Limiting

In addition to detecting DDoS attacks, rate limiting is another component of the proposed scheme. In our approach, if a flow f is classified as attack, rate limiting technique will be applied to the flow in order to limit malicious traffic to a manageable level. One important design principle of our rate limiting strategy is that the rate limit applied to a malicious flow must reflect to current behavior of the flow. In this way, we can further restrict an ill-behaviored flow when it continually violates the legitimate flow model. From this point of view, the α value, which represents the malicious level of the monitored flow, serves as a rate limiting parameter. For the first time a flow is classified as an attack flow, the correspondent rate limit is:

$$rl = R * (1 - \alpha) \tag{10.2}$$

In Eq. 10.2, rl denotes the rate limit and R denotes the sending rate of the monitored for in the current observation interval. In the following observation periods, if the malicious flow does not show compliance to the legitimate flow model, it will be restrict further, according to the following formula:

$$rl_{new} = \min(rl_{old}, R) * (1 - \alpha) * \frac{P_s}{P_s + P_{drop}} \tag{10.3}$$

In Eq. 10.3, rl_{new} denotes a new rate limit to be applied on the malicious flow. rl_{old} denotes the rate-limit applied on the flow in previous observation interval. R represents the realized sending rate in previous observation interval. P_s is the total number of packets sent in the flow and P_{drop} is the total number of packets dropped because of the imposed rate limit.

In this way, flows that are part of DDoS attacks would be quickly restricted to a very low rate since the attack scripts would persistently send attack packets to the victim. Consequently, the fraction $(P_s)/(P_s+P_{drop})$ would become very low quickly.

Next, consider the case that the rate limited flow is mis-classified. In this case, TCP-based network applications will stop sending packets when the network is highly congested. Since the TCP/IP protocol will actively slow down the sending rate, the flow will show compliance with the legitimate flow model. In our approach, whenever an attack flow is compliant with the normal flow model for consecutive *PenaltyPeriod* observation periods, the flow is considered a transient flow and the recovery process begins. In the recovery process, rate limit are carefully removed according to the following equation:

$$rl_{new} = rl_{old} * \frac{1}{\alpha} * \frac{P_s}{P_s + P_{drop}} \tag{10.4}$$

In Eq. 10.4, it is clear that the speed of recovery is controlled by α and $P_s/(P_s + P_{drop})$. Both reflect the current behavior of the monitored flow. When the rate limit reaches *MaxRate*, a transient flow is classified as a normal flow, and rate limit is completely removed.

4. Performance Evaluation

To evaluate the performance of the proposed scheme, we implemented both prototypes of D-WARD and our approach on a machine which runs the FreeBSD operating system. In our experiment, two types of DDoS attacks are conducted: TCP SYN flooding attack and link overloading attack. In the TCP SYN flooding attack, each attack agent floods the victim with TCP SYN packet at the maximum rate of 100KBps. In this experiment, we will show that attacks detected by D-WARD can also be detected by our approach. Even further, our scheme can detect the attacks earlier than D-WARD. Next, In the link overloading attack, each agents sends the victim at the maximum rate of 100KBps. The link bandwidth of the victim is restricted to 500KBps. This is accomplished by using Dummynet [Rizzo, 1997]. (there are in total 10 agents) In this experiment, we will show that our approach can detect attacks which cannot detected by D-WARD. For both types of attacks, we replicate the four attack scenarios tested in D-WARD. That is, constant rate attack , pulsing attack , increasing rate attack and gradual pulsing attack.

4.1 Experimental Results

Fig. 10.4, 10.5, 10.6, and 10.7 show the experimental results of TCP SYN attack. The x-axis denotes time measured in second and the y-axis stands for attack bandwidth measured in KB per second. The line with "x" symbols represents the attack bandwidth generated by attack agents. The line with triangle symbols represents attack bandwidth going through D-WARD, and the line with square symbols denotes the attack bandwidth passing by the proposed scheme. According to the figure, our scheme can detect the attack earlier than D-WARD. This makes our scheme more effective than D-WARD mainly because the thresholds used in our scheme are continually adjusted and derived from the past behavior of the monitored flows.

Next, we examine the experimental results of link overloading attacks. In this experiment, by controlling the attack sending rate, the O/I value of the attack flow only reaches 2, smaller than threshold value 3 used in D-WARD.

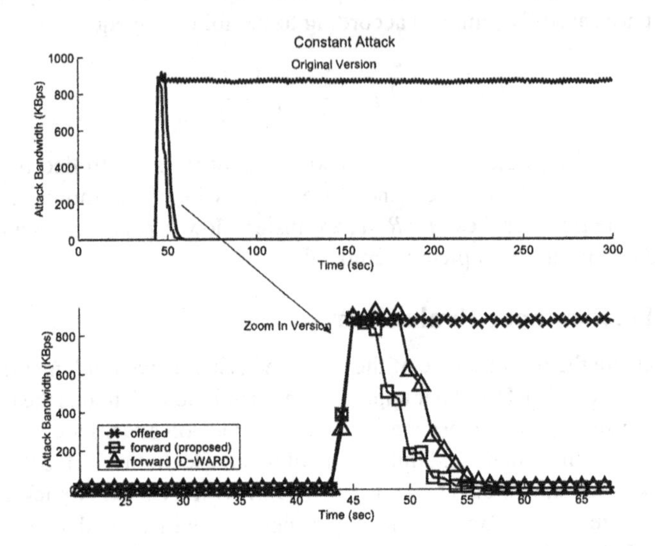

Figure 10.4. Constant SYNC attack.

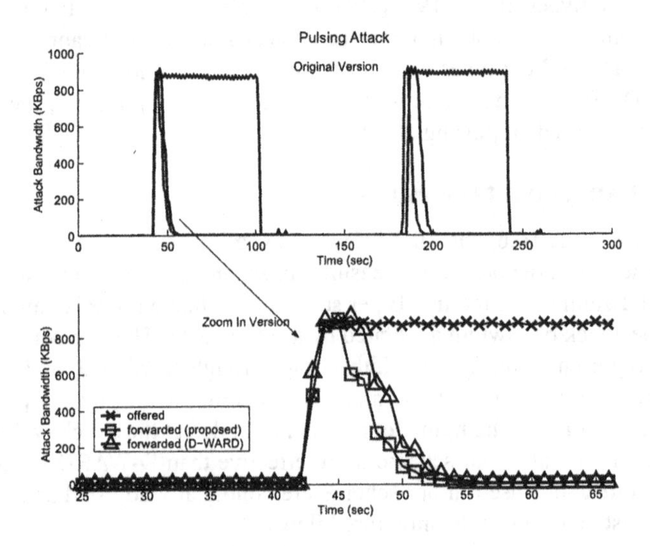

Figure 10.5. Pulsing SYNC attack.

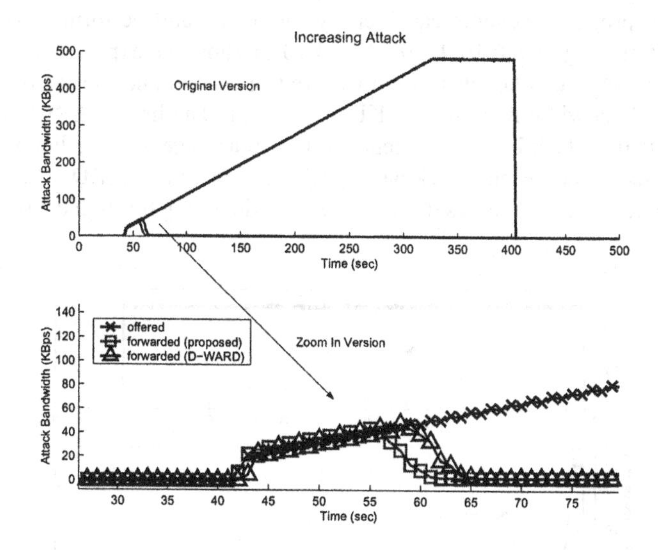

Figure 10.6. Increasing SYNC attack.

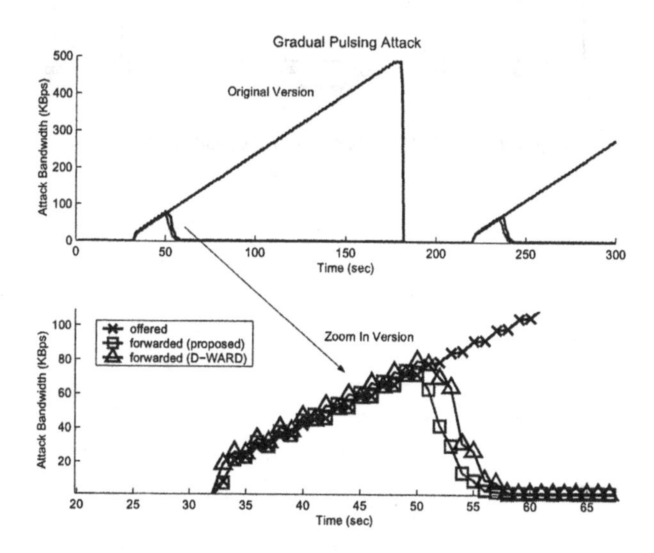

Figure 10.7. Gradual SYNC attack.

Thus, D-WARD is unable to detect the presence of the attack. On the other hand, the proposed scheme can identify the attack and perform subsequent rate limiting. Fig. 10.8, 10.9, 10.10, and 10.11 show the experimental result. Similarly, the x-axis denotes time measured in second and the y-axis stands for attack bandwidth measured in KB per second. The line with "x" symbols represents the attack bandwidth generated by attack agents. The line with triangle symbols represents attack bandwidth passing by D-WARD, and the line with square symbols denotes the attack bandwidth going through the proposed scheme.

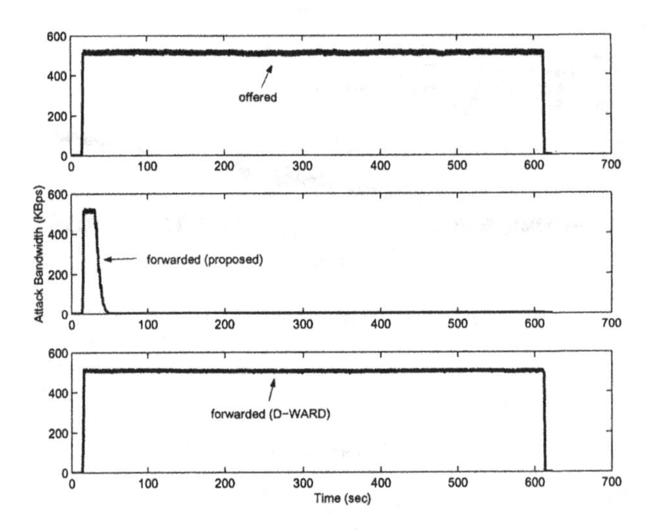

Figure 10.8. Constant bandwidth overloading attack.

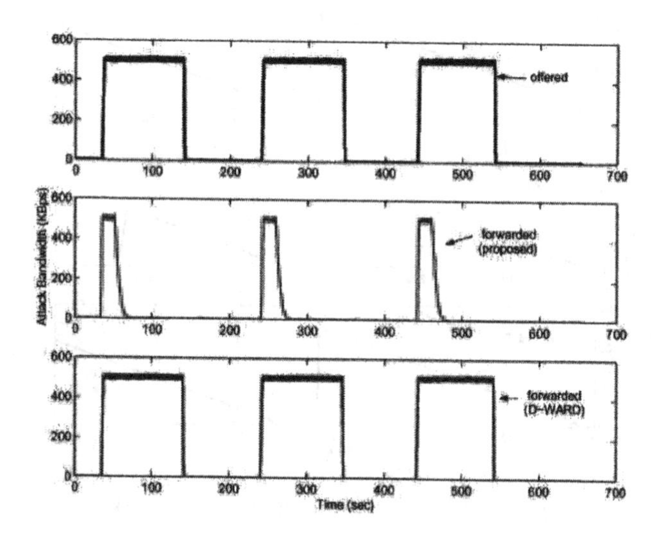

Figure 10.9. Pulsing bandwidth overloading attack.

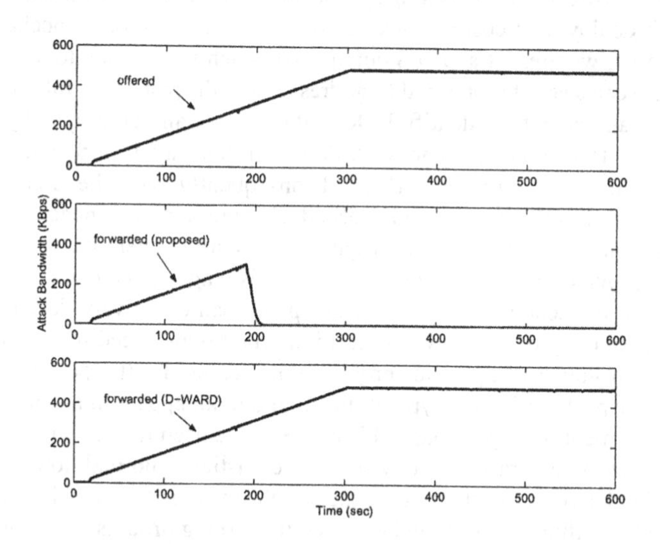

Figure 10.10. Increasing bandwidth overloading attack.

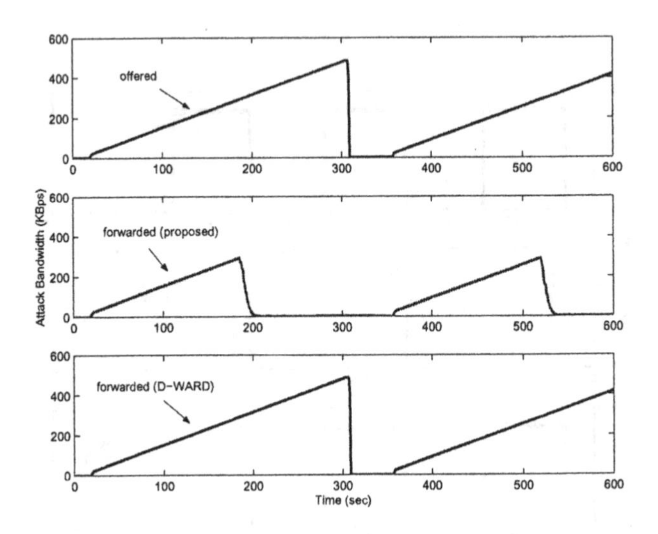

Figure 10.11. Gradual bandwidth overloading attack.

5. Conclusion and Future Work

Technology resisting DDoS attacks has drawn considerable attention in recent years. However, most existing approaches suffer from either low detection rate, high deployment cost, or lack of effective attack response mechanisms. In this paper, we present a DDoS defense approach which monitors two-way traffic between a set of monitored IP addresses and the rest of the Internet. Our approach can accurately identify DDoS attack flows and consequently apply rate-limiting to the malicious network flows. In this way, DDoS attack traffic can be contained in source networks, and consequently lower the effectiveness of the attack. To effectively stop DDoS attacks, our approach needs to be deployed in routers serving as default gateways. With cooperative routers, our approach provides an effective defense mechanism against DDoS attacks.

Although the scheme presented in this paper can effectively detect DDoS attacks based on two-way flows, several important issues need further investigation. For instance, one pressing problem not addressed in this paper is how to establish the profile of a new type of flow that did not appear in historical traffic data. As mentioned previously, historical traffic used in the learning phase must not have attack traffic; otherwise, characteristics of normal flow behavior may not be derived. To achieve this, the simplest way is to manually examine the collected traffic before it can be passed to learning process. However, it is clear that this approach is not efficient since it requires an extensive amount of time to examine the traffic manually. Furthermore, investigation for effective creation of new flow profiles is desirable. Additionally, an effective profile management system is important and critical to the overall performance of the DDoS defense system. With all the systems putting together, the source-end DDoS defense can be quite effective and consequently deter DDoS attacks.

References

[man,] MANAnet DDoS White Papers. http://www.cs3-inc.com/mananet.html.

[Net,] NetRanger Overview. http://www.cisco.com/univercd/cc/td/doc /product/iaabu/csids/csids1/csidsug/overview.htm.

[NFR,] NFR Network Intrusion Detection. http://www.nfs.com/products/NID/.

[Chang, 2002] Chang, K. C. (2002). Defending against Flooding-Based Distributed Denial-of-Service Attacks: A Tutorial. In *IEEE Communications Magazine*, volume 40, pages 42–51.

[Dean et al., 2002] Dean, Drew, Franklin, Matt, and Stubblefield, Adam (2002). An Algebraic Approach to IP Traceback. *ACM Transactions on Information and System Security*, (2):119–137.

[Feinstein et al., 2003] Feinstein, L., Schnackenberg, D., Balupari, R., and Kindred, D. (2003). Statistical Approaches to DDoS Attack Detection and Response. In *Proceedings of DARPA Information Survivability Conference and Exposition*, volume 1, pages 303–314.

[Ferguson, 1998] Ferguson, P. (1998). Network Ingress Filtering: Defending Denial of Service Attacks Which Employ IP Source Address Spoofing.

[Ioannidis and Bellovin, 2002] Ioannidis, J. and Bellovin, S. M. (2002). Implementing Pushback: Router-Based Defense Against DDoS Attacks. In *Proceedings of Networks and Distributed System Security Symposium*.

[Juels and Brainard, 1999] Juels, A. and Brainard, J. (1999). Client puzzles: A cryptographic countermeasure against connection depletion attacks. In *Proceedings of Networks and Distributed System Security Symposium*, pages 151–165.

[Leiwo et al., 2000] Leiwo, J., Nikander, P., and Aura, T. (2000). Towards network denial of service resistant protocols. In *Proceedings of 15th International Information Security Conference*, pages 301–310.

[Mahajan et al., 2002] Mahajan, R., Bellovin, S., Floyd, S., Paxson, V., and Shenker, S. (2002). Controlling high bandwidth aggregates in the network. *ACM Computer Communications Review , 32(3)*, pages 62–73.

[Mann et al., 2000] Mann, G. R., Watson, D., Jahanian, F., and howell, P. (2000). Transport and Application Protocol Scrubbing. In *Proceedings of INFOCOM*, pages 1381–1390.

[Mirkovic et al., 2002a] Mirkovic, J., Martin, J., and Reiher, P. (2002a). Taxonomy of DDoS Attacks and DDoS Defense Mechanisms. Technical Report 020018, UCLA Technical.

[Mirkovic et al., 2002b] Mirkovic, J., Prier, G., and Reiher, P. (2002b). Attacking DDoS at the Source. In *Proceedings of International Conference on Network Protocols*, pages 312–321.

[Moore et al., 2001] Moore, D., Voelker, G., and Savage, S. (2001). Inferring internet denial-of-service activity. In *Proceedings of 10th USENIX Security Symposium*.

[Park and Lee, 2001] Park, K. and Lee, H. (2001). On the Effectiveness of Router-Based Packet Filtering for Distributed DoS Attack prevention in Power-Law Internets. In *Proceedings of ACM Sigcomm*, pages 15–26.

[Rizzo, 1997] Rizzo, Luigi (1997). Dummynet: a simple approach to the evaluation of network protocols. *ACM Computer Communication Review*.

[Roesch, 1999] Roesch, Martin (1999). Snort - Lightweight Intrusion Detection for Networks. In *Proceedings of LISA'99: 13th Systems Administration Conference*, pages 229–238.

[Savage et al., 2001] Savage, Stefan, Wetherall, David, Karlin, Anna, and Aderson, Tom (2001). Network Support for IP Traceback. *IEEE/ACM Transactions on Networking*, (3):226–237.

[Savage et al., 2000] Savage, Stefan, Wetherall, David, Karlin, Anna R., and Anderson, Tom (2000). Practical Network Support for IP Traceback. In *Proceedings of SIGCOMM Conference*, pages 295–306.

[Shaprio and Hardy, 2002] Shaprio, J. and Hardy, N. (2002). EROS: A principle-driven operating system from the ground up. *IEEE Software*, pages 26–33.

[Song and Perrig, 2001] Song, Dawn and Perrig, Adrian (2001). Advanced and Authenticated Marking Schemes for IP Traceback. In *Proceedings of IEEE INFOCOM Conference*, pages 878–886.

[Sung and X, 2002] Sung, M. and X, J. (2002). IP Traceback-Based Intelligent Packet Filtering: A Novel Technique for Defending against Internet DDoS Attacks. In *Proceedings of International Conference on Network Protocols*, pages 302–311.

[T. Aura and Leiwo, 2001] T. Aura, P. Nikander and Leiwo, J. (2001). DOS-Resistant Authentication with Client Puzzles. *Lecture Notes in Computer Science*, 2133.

[Wang and Reiter, 2003] Wang, X. and Reiter, M. (2003). Defending Against Denial-of-Service Attacks with Puzzle Auctions. In *Proceedings of IEEE Symposium on Security and Privacy*, pages 78–92.

Chapter 11

BEAGLE: TRACKING SYSTEM FAILURES FOR REPRODUCING SECURITY FAULTS

Chang-Hsien Tsai
Department of Computer Science and Information Engineering
National Chiao Tung University

Shih-Hung Liu
Department of Computer Science and Information Engineering
National Chiao Tung University

Shuen-Wen Huang
Institute of Information Science Academia Sinica

Shih-Kun Huang
Department of Computer Science and Information Engineering
National Chiao Tung University,
and Institute of Information Science Academia Sinica

Deron Liang
Institute of Information Science Academia Sinica,
Department of Computer Science
National Taiwan Ocean University

Abstract Software vulnerabilities can be attributed to inherent bugs in the system. Several types of bugs introduce faults for not conforming to system specifications and failures, including crash, hang, and panic. In our work, we exploit security faults due to crash-type failures. It is difficult to reconstruct system failures after a program has crashed. Much research work has been focused on detecting program errors and identifying their root causes either by static analysis or observing their running behavior through dynamic program instrument. Our goal is to design a tool that helps isolate bugs. This tool is called BEAGLE (Bug-tracking by Execution Auditing from Generated Logs and Errors). BEAGLE periodically makes stack checkpoints of program in execution. If the software crashes, we can approximate to the latest checkpoint and infer the precise corrupt site. After identifying the site of control state corruption, tainted input analysis will determine system exploitability if untouched passed through the corrupt site. Several case studies of corrupt site detections and tainted input analysis prove the applicability of our tool.

Keywords: Dynamic Analysis, Software Wrapper , COTS Vulnerability Testing, Control State Corruption

1. Introduction

Abnormal program running behaviors have much to do with software security, including access race conditions, tainted input attributed to system failures such as buffer overflow and indefinitely hang of denial of service. We can classify such cases of anomaly into crash, hang, and panic. Especially, crashed software may be exploited to be a potential vulnerability. It is difficult to reconstruct system failures after a program has crashed due to corrupted control state and gaps between crash sites and corrupt sites.

In order to meet the time to market, software releases with unintended flaws. Some of them cause software crash, while others may introduce security vulnerabilities. Our goal is to design a tool that helps analyze the program running behavior and determine if it is an exploitable vulnerability. We try to intercept and monitor running behaviors during programs in execution when only COTS (Commercial Off The Shelf) executables available for analysis.

We develop a run-time instrument and interception tool called BEAGLE to periodically monitor software running behavior. If the software crashes, we can approximate to the latest checkpoint and determine if the failure point is exploitable through tainted input analysis. We can observe the internal behavior of running programs, such as API call sequence, call parameters and return values through wrapping system call API techniques, and determine whether these things are anomalous or not. We develop such a dynamic instrument tool able to determine if the crash site is security exploitable.

We investigate the design and implementation of the BEAGLE system to instrument the interfaces between the software application and the operating system functions with an interactive software wrapper. This wrapper cannot only intercept the functions to record the parameters and the return value but also receiving testing directives to replace calling parameters and the return value with any arbitrary value. We could use this tool to easily instrument the application, change the intended OS function call parameters with testing data and observe the response of the application to find out the suspicious crash sites.

2. The Detection of Control State Corruption

Our research uses the following approaches to manifest and analyze the crash process as precisely as possible.

2.1 Distinguishing Corrupt site and Crash site

We must clarify the difference between corrupt site and crash site. The crash site is obviously the point where software crashes. The corrupt site is the point where software stack is corrupted. For example, in Figure 11.1, function *foo* passes its local buffer *buf* to function *bar*, which overflows the buffer. After `strcpy()` returns, the stack is corrupted. However, the program do not crash until the function foo returns(in line 4). Therefore, we must monitor the corrupt site to find the root causes.

2.2 Algorithm

$< ebp_1, ebp_2, \ldots, ebp_{n-1}, ebp_n >$ is a strictly decreasing list. ebp_i points to the address of ebp_{i-1}, $1 < i \leq n$. When entering a function ebp_{n+1} is appended to the list; when exiting a function ebp_n is removed from the list.

Initial time, saved EBP of startup function, $ebp_s := ebp_1$; saved EBP of main function $ebp_m := ebp_2$. At each checkpoint, Beagle performs the following corrupt site detection algorithm:

Corrupt_Site_Detection { traverse from ebp_n to the head of list if $ebp_s \neq ebp_1$ AND $ebp_m \neq ebp_2$ then *the stack is corrupt* else if $ebp_s = ebp_1$ AND $ebp_m \neq ebp_2$ then *in exception handler* }

```
1. void foo(void){
2.     char buf[8];
3.     bar(buf);
4. }      /*crash site*/
5. void bar(char *buf){
6.     strcpy(buf, "this is a long string");
       /*corrupt site*/
7.         ...
8.}
```

Figure 11.1. A program with buffer overflow.

2.3 API call interception

API call interception technique is the groundwork of this work. The ability to control API function calls is extremely helpful and enables developers to track down the internal actions happening during the API call. The purpose of API call interception is to take control of some execution code. That is the so-called "stub" to force the target application to execute the injected code. Therefore, the injected code can easily monitor the program by parameter logging, return value checking, stack dump, EBP tracing, etc. According to the ways of injection of users' DLL into the target process and interception mechanisms, there exists some different kind of works for different purposes. After the consideration about stack frame evolving due to added monitor function and the completeness of the API interception mechanism, Detours [Galen and Doug, 1999] is chosen to be the framework of this work.

2.4 Function call wrapping

For purpose of monitoring the stack frame evolving and tracing EBP and return pair, however, API call interception still seems too coarse to pinpoint the reason why the application programs crash. Actually, the most ideal scenario for crash analysis is to figure out which line of code is the onset of bugs, and it is impossible without source code. What we can do furthermore is to wrap user functions to achieve the finer-grained monitoring. The process of function call wrapper generation is shown in Figure 11.2. Function call wrapping is especially helpful to catch the site resulting in crashes happening on the stack. For instance, if a function in a program does some string manipulation without careful bound checking, it may crash when the string in process is out of bound. Such vulnerabilities bring about the classic and simple attack, i.e. stack overflow. By overwriting the return address through stack variables overflowing, the attacker can intercept the programs when this function returns. Therefore, the control jumps to a location where the attacker would have inserted malicious code. To deserve to be mentioned, the buffer overflow attack is a kind of injection/interception mechanism. Compared with the API interception techniques mentioned above, buffer overflow cannot successfully return back to the correct site after some destructive activities since the return address and the stack is overwritten. The principle of function wrapping is similar to what Detours does in the API call interception. Detours replaces the first few instructions of the target API with unconditional jump to the user-provided monitor function. The primary difference between Detours and this function wrapper is as follows:

1 Detours acquires the API call entry address from static linking. However, this function wrapper acquires the user function from the disas-

sembly of the binary code of the application program through the FREE tool named *OllyDBG*.

2 Detours only instruments the prologue of the API call. However, this function wrapper instruments both the prologue and epilogue of the user function. Comparing the stack tracing in a function's entry and exit is extremely helpful to detect the anomaly of the stack.

2.5 Input pollutant tracing

Establishing the bridge connecting the software robustness and security is a brand-new and fantastic idea in the research area of software testing. Traditional testing techniques are well equipped to find the bugs that violate the specification, but lack of looking for how these bugs relate to the security issues. For example, there are plenty of application crashes during our everyday life and you may wonder whether bugs leading to these crashes are security-related.

When stack corruption occurs, current stack frame is infected by pollutants. We apply the LCS(Longest Common Subsequence) in the following algorithm to correlate the relevant local variables. If the pollutant variables are related to some input, the bug may become security vulnerability.

---

input_pollutant_tracing { for $i = n$ downto 2 { $lcs\_len = LCS([ebp_{n-1}, ebp_n], pollutant)$; if ($lcs\_len > threshold$) then correlate the local variables } }

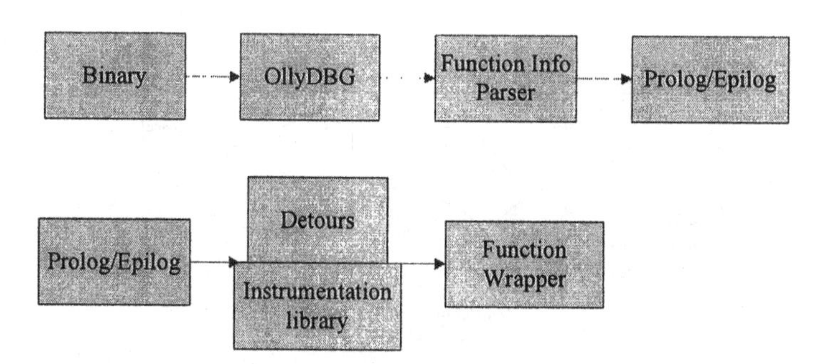

Figure 11.2. Process of the function wrapper generation

3. The BEAGLE System Design and Implementation

The instrument tool mentioned above could help testers to know why the programs crash by observing the stack and input tracing. Furthermore, to manifest the exploit process of the known vulnerable programs is another proof that this tool is useful. Using the log of runtime monitoring on the running applications, we hope that this tool could help analyze why this software could be exploited. There is an instrument tool to communicate with the API/function wrapper DLL that is injected into the target process. During the execution of the application program, testers may want to modify the parameter or return values of a certain suspicious functions. Figure 11.3 shows the system architecture.

3.1 Binary Disassembly

Our system relies heavily on the disassembly ability of OllyDBG, which is a 32-bit assembler debugger on Microsoft Windows. It does much work on binary code analysis that we could utilize especially when the source is not available. It could recognize procedures, API calls, and complex code constructs, like call to jump to procedure. These analyses help us parse the disassembly of the application to retrieve the necessary information such as procedure call sitse, and entry addresses. In addition, it could disassemble all the executable modules the application loads.

3.2 Function Info Parser

In order to transfer control from the execution of the application process to our runtime-generated stub, we need to replace instructions at the function prologue and epilogue with a `JMP` to the stub. The type of the procedures

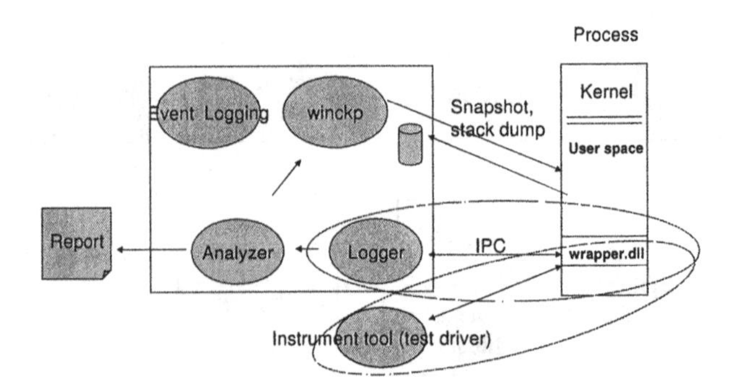

Figure 11.3. The architecture of Beagle

we recognize is the typical function prologue and epilogue, which will do operations on the stack and frame pointer. Our function info parser retrieves prologue/epilogue information that is needed by the instrument library .

When doing instrument of the prologue, the targeted library will check the length of the binary to be replaced. If the length is less than 5 bytes, it means that there is not enough space to substitute the prologue for the `JMP` instruction and we leave this kind of procedure to breakpoint interruption instruction instrumentation if needed. There may be multiple return sites of this function, but not all of them have enough space to be instrumented.

3.3 Instrumentation Library

We develop an instrumentation library to replace the certain functions at runtime. According to the information provided by the function info parser, the instrument library will allocate the space for the stub and append the intended instructions on the stub. The most important instruction is to CALL the monitor function where we could backtrace the stack for corruption detection. Detours provides some useful library to append the certain instruction on the stub.

4. Experiments and Assessment

4.1 Calibration of Checkpoint Interval

We checkpoint *notepad.exe* with different intervals as shown in Figure 11.4. With finer-grained interval, we get closer picture of the stack trace. However, due to the hardware limitation and OS interrupt, the length of the interval is limited.

4.2 Stack Tracing and Tainted Input Analysis

To validate the correctness of the BEAGLE prototype, we need to verify that our stack corrupt site detection does point out the vulnerable function where the stack is polluted. We instrument RobotFTP Server 1.0, which has a known stack overflow vulnerability, to demonstrate that BEAGLE could detect the abnormal stack at runtime when running the exploit and terminate the program. The description of the vulnerable program follows. RobotFTP Server is an FTP server for the Microsoft Windows platform. It has a non-trivial buffer overrun bug in the function that processes the login information that an FTP client sends. An attacker can first login with a username longer than 48 characters and login again with a username 1994 character long to overflow the return address of this function. When this program is running under the BEAGLE instrumentation, this buffer overflow will be detected and terminate the program to prevent from transferring control to the attacker's payload. The result is shown as Figure 11.5.

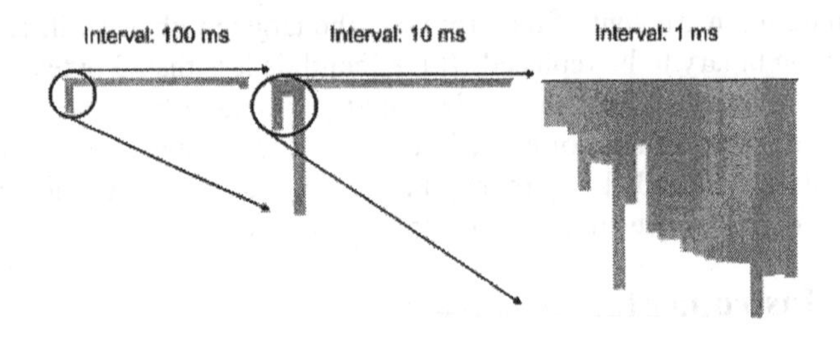

Figure 11.4. Calibration of the notepad.exe stack trace

```
DebugView on \\BEAGLE (local)
File Edit Capture Options Computer Help
Debug Print
[2004] Stack top: 014E0000 Stack base: 014DD000
[2004] --[51e994][TID:3bc] 14def30 437460 14dfebc 435d9a 14dfed4 4351cb 14dffa4 522824 14dff
[2004] Stack top: 014E0000 Stack base: 014DD000
[2004] --[437420][TID:3bc] 14dfebc 435d9a 14dfed4 4351cb 14dffa4 522824 14dffb4 52285f 14dff
[2004] Stack top: 014E0000 Stack base: 014DD000
[2004] ++[4394ac][TID:3bc] 14dfebc 435db4 14dfed4 4351cb 14dffa4 522824 14dffb4 52285f 14dff
[2004] Stack top: 014E0000 Stack base: 014DD000
[2004] --[4394ac][TID:3bc] 14dfebc 435db4 14dfed4 4351cb 14dffa4 522824 14dffb4 52285f 14dff
[2004] Stack top: 014E0000 Stack base: 014DD000
[2004] ++[4339c8][TID:3bc] 14dfebc 435df2 14dfed4 4351cb 14dffa4 522824 14dffb4 52285f 14dff
[2004] Stack top: 014E0000 Stack base: 014DD000
[2004] --[4339c8][TID:3bc] 14dfebc 435df2 14dfed4 4351cb 14dffa4 522824 14dffb4 52285f 14dff
[2004] Stack top: 014E0000 Stack base: 014DD000
[2004] ++[439574][TID:3bc] 14dfebc 435c7f 14dfed4 4351cb 14dffa4 522824 14dffb4 52285f 14dff
[2004] Stack top: 014E0000 Stack base: 014DD000
[2004] DEBUG: case3
[2004] --[439574][TID:3bc] 41414141 58585858 14dfed4 4351cb 14dffa4 522824 14dffb4 52285f 14d
[2004] ++++[439574][TID:3bc] 14dfebc 435c7f 14dfed4 4351cb 14dffa4 522824 14dffb4 52285f 14d
[2004] [Corrupted]Stack is corrupted in this function. (439574) 4 records
```

Figure 11.5. The stack backtrace of the RobotFTP Server 1.0 with overlong input

We can see first frame pointer and return address pair in the third line from bottom, (41414141, 58585858), and this is the second overlong input username. Before program returns from this vulnerable function, our epilogue monitor function backtraces the stack and discovers that this stack trace is abnormal by comparing the stack trace in the prologue monitor function.

5. Related Work

A considerable amount of work has been performed on detecting program errors and identifying their root causes either by static analysis or observing their running behavior through dynamic program instrumentation. In this section we review different work in each category and relate them to our work.

5.1 Static analysis

Static analysis is based on the information provided by the source code. It may validate the call sequence to find the program error or check if the vulnerable function call is used without actually executing the application. The drawbacks of this way to find bugs are:

1 there are too many program states to verify.

2 it cannot know some dynamic information such as pointers and should use other inexact measures to analyze [Guyer and Lin, 2003, Shapiro and Horwitz, 1997, Bjarne, 1996].

Our method combines the static analysis of binary disassembly and dynamic instrument and monitor to point out where the corruption occurs.

Chen and Wagner use a formal approach to examine whether the program violates the pre-defined security properties, which are described by Finite State Automata (FSA) [Chen and Wagner, 2002] . The programs to be tested are modeled as pushdown automation (PDA) and MOPS uses model-checking techniques to determine the reachability of exception states in the PDA. Liblit and Aiken present an algorithm for computing a set of paths given a crash site and a global control flow graph [Liblit and Aiken, 2002]. Furthermore, it uses some post-crash artifacts such as the stack trace and the event trace to reduce the set of possible execution paths.

5.2 System Call Interception Techniques

System call interception is the fundamental technique in our work. Detours developed by Hunt and Brubacher is a library for instrument of arbitrary Win32 functions on x86 machines [Galen and Doug, 1999]. It replaces the first few instructions of the target function with unconditional jump, which points to the user-provided detour function. Users can do the interception work in the corresponding detour function.

Pietrek develops API-SPY in his book [Pietrek, 1995]. API-SPY tools lists API's name in the order they are called, and record the parameters as well as the return value. The purpose of this work, the same as Detours, is to get control before the intended target function call is reached. However, the technique used in API-SPY is DLL redirection by modifying the Import Address Table (IAT) , much different from the way Detours used, which is to modify the target function's prologue code to transfer control by inserting a JMP instruction at the start of the function.

5.3 Tracking Down Software Bugs Using Runtime Inspection

A few systems automatically add codes in the source and observe the behavior of these codes at runtime. The difference from our work is that we instrument the runtime process image, not the source. Therefore even if we don't have the source, we can still detect program errors and add survival patches. Hangal and Lam present DIDUCE for tracking down software bugs using automatic anomaly detection [Hangal and Lam, 2002]. DIDUCE aids programmers in detecting complex program errors and identifying their root causes. It dynamically formulates hypotheses of invariants obeyed by the program. Our work is also based on runtime inspection, but in different method from DIDUCE. DIDUCE observes the invariants at runtime and check if the program violates. Binary Rewriting protects the integrity of the return address on the stack by modifying the binary code [Manish and Chiueh, 2003]. The difference from our work is that its detection on stack overflow has false positive when the corruption occurs not in the current stack frame.

5.4 Fault Triggering and Robustness Testing

Fault triggering systems aide in producing system crashes and we can examine whether these crash are exploitable or not. Ghosh and Schmid present an approach to testing COTS software for robustness to operating system exceptions and errors [Ghosh and Matthew, 1999]. That is, they bring up an idea to assess the robustness of Win32 applications. It instruments the interface between the software application and the Win32 APIs. By manipulating the APIs to throw exceptions or return error codes, it analyzes the robustness of the application under the stressful conditions. Whittaker and Jorgensen summarize the experiences of breaking software in their lab [James and Jorgensen, 1999]. By studying how these software failed, they presents four classes of software failures: improperly constrained input, improperly constrained stored data, improperly constrained computation and improperly constrained output. Software testers can use the four classes of failures to break the software.

6. Conclusions

We have tried to build up relationship between system robustness and software security. Unreliable software with inherent bugs may be exploited to violate security specifications, meant to be security faults. Types of bugs are either faults not conforming to system specifications or failures such as crash, hang, and panic. We design and implement the BEAGLE system to back-track crash type failures and analyze tainted input by an input pollutant tracing algorithm to determine if such failures are security exploitable. Crash-type failures are potentially vulnerable to be exploited by tainted input due to corruption of control state. We try to approximate sites of control state corruption by stack checkpoints and monitoring run-time status. Known exploits have been tested to proof the applicability of the system. We hope to discover more failures that will introduce security exploits with finer-grained stack checkpoints and further improve the precision of approximation process for corruption detection.

References

[Chen and Wagner, 2002] Chen, Hao and Wagner, David (2002). MOPS: an infrastructure for examining security properties of software. In Atlury, Vijay, editor, Proceedings of the 9th ACM Conference on Computer and Communication Security (CCS-02), pages 235-244, New York. ACM Press.

[Ghosh and Matthew, 1999] Ghosh, Anup K. and Schmid, Matthew (1999). An approach to testing cots software for robustness to operating system exceptions and errors. In Proceedings of the 10th International Symposium on Software Reliability Engineering.

[Guyer and Lin, 2003] Guyer, Samuel Z. and Lin., Calvin (2003). Client-driven pointer analysis. In Proceedings of the 10th International Static Analysis Symposium, pages 214-236.

[Hangal and Lam, 2002] Hangal, Sudheendra and Lam, Monica S. (2002). Tracking down software bugs using automatic anomaly detection. In Proceedings of the 24th International Conference on Software Engineering (ICSE-02), pages 291-301, New York. ACM Press.

[Galen and Doug, 1999] Hunt, Galen and Brubacher, Doug (1999). Detours: Binary interception of Win32 functions. In Proceedings of the 3rd USENIX Windows NT Symposium (WIN-NT-99), pages 135-144, Berkeley, CA. USENIX Association.

[Liblit and Aiken, 2002] Liblit, Ben and Aiken, Alex (2002). Building a better backtrace: Techniques for postmortem program analysis. Technical Report CSD-02-1203, University of California, Berkeley.

[Pietrek, 1995] Pietrek, Matt (1995). Windows 95 System Programming Secrets. IDG Books.

[Manish and Chiueh, 2003] Prasad, Manish and cker Chiueh, Tzi (2003). A binary rewriting defense against stack based overflow attacks. In Proceedings of the USENIX Annual Technical Conference, pages 211-224.

[Shapiro and Horwitz, 1997] Shapiro, Marc and Horwitz, Susan (1997). Fast and accurate flow-insensitive points-to analysis. In Proceedings of the 24th ACM SIGPLAN-SIGACT symposium on Principles of programming languages, pages 1-14. ACM Press.

[Bjarne, 1996] Steensgaard, Bjarne (1996). Points-to analysis in almost linear time. In Proceedings of the 23rd ACM SIGPLAN-SIGACT symposium on Principles of programming languages, pages 32-41. ACM Press.

[James and Jorgensen, 1999] Whittaker, James A. and Jorgensen, Alan (1999). Why softwarre fails. SIGSOFT Software Engineering Notes, 24(4):81-83.

IV

MULTIMEDIA SECURITY

Chapter 12

WEB APPLICATION SECURITY—PAST, PRESENT, AND FUTURE *

Yao-Wen Huang[1,2] and D. T. Lee[1,3]
{ywhuang,dtlee}@iis.sinica.edu.tw

[1] *Institute of Information Science, Academia Sinica*
Nankang 115, Taipei, Taiwan

[2] *Department of Electrical Engineering,*
National Taiwan University

[3] *Department of Computer Science and Information Engineering,*
National Taiwan University

Abstract Web application security remains a major roadblock to universal acceptance of the Web for many kinds of online transactions, especially since the recent sharp increase in remotely exploitable vulnerabilities has been attributed to Web application bugs. In software engineering, software testing is an established and well-researched process for improving software quality. Recently, formal verification tools have also shown success in discovering vulnerabilities in C programs. In this chapter we shall discuss how to apply software testing and verification algorithms to Web applications and improve their security attributes. Two of the most common Web application vulnerabilities that are known to date are script injection , e.g., SQL injection, and cross-site scripting (XSS) . We will formalize these vulnerabilities as problems related to information flow security—a conventional topic in security research. Using this formalization, we then present two tools, WAVES (Web Application Vulnerability and Error Scanner) and Web-SSARI (Web Application Security via Static Analysis and Runtime Inspection) , which respectively utilize software testing and verification to deal in particular with script injection and XSS and address in general the Web application security problems. Finally we will present some results obtained by applying

*This work was supported in part by the National Science Council under the Grants NSC-93-2213-E-001-013, NSC-93-2422-H-001-0001, and NSC-93-2752-E-002-005-PAE.

these tools to real-world Web applications that are in use today, and give some suggestions about the future research direction in this area.

Keywords: Web application security, software testing, software verification

1. Introduction

As World Wide Web usage expands to cover a greater number of B2B (business-to-business), B2C (business-to-client), healthcare, and e-government services, the reliability and security of Web applications has become an increasingly important concern. In a Symantec analysis report of network-based attacks, known vulnerabilities, and malicious code recorded throughout 2003 [Higgins et al., 2003], eight of the top ten attacks were associated with Web applications; the report also stated that port 80 was the most frequently attacked TCP port. In addition to holding Web applications responsible for the sharp increase in moderately severe vulnerabilities found in 2003, the authors of the report also suggested that Web application vulnerabilities were by far the easiest to exploit.

Web application insecurity is attributed to several factors. Firstly, the Web, which was initially designed as a data-delivery platform, has quickly evolved into a complex application platform on top of which more and more sophisticated applications have been developed. As a result, Web specifications have grown rapidly to meet rising demands, and browsers and Web-development languages fought a "feature war" to win market share. Unfortunately, security issues have been left as an afterthought. The fast-expanded features did help Web growth; however, many security side effects they induced have become today's major concern for Web adoption. Secondly, since software vendors are becoming more adept at writing secure code and developing and distributing patches to counter traditional forms of attack (e.g., buffer overflows), hackers are increasingly targeting Web applications. Web application vulnerabilities are hard to eliminate because most Web applications a) go through rapid development phases with extremely short turnaround time, and b) are developed in-house by corporate MIS engineers, most of whom have less training and experience in secure software development compared to engineers at IBM, Sun, Microsoft, and other large software firms. Lastly, current technologies such as anti-virus software and network firewalls offer comparatively secure protection at the host and network levels, but not at the application level [Curphey et al., 2002]. When network and host-level entry points are relatively secure, the public interfaces to Web applications become the focus of attacks [Meier et al., 2003] [Curphey et al., 2002]. In this chapter we shall first provide a brief description of common Web application vulnerabilities. We then describe possible automated approaches to eliminating such vulnerabilities. Finally we will give some concluding remarks and present a few possible avenues for future work in this area.

2. Common Web Application Vulnerabilities

Two of the most common Web application vulnerabilities are script injection (e.g., SQL injection) and cross-site scripting (XSS). In this section, we will briefly describe these vulnerabilities; the reader is referred to Scott and Sharp [Scott and Sharp, 2002a] [Scott and Sharp, 2002b], Curphey et al. [Curphey et al., 2002], and Meier et al. [Meier et al., 2003] for more details.

2.1 Cross-site Scripting

On Feb 2, 2000, CERT Coordination Center issued an advisory [CERT, 2001] on "cross-site scripting" (XSS) attacks on Web applications. This hard-to-eliminate threat soon drew the attention and spawned active discussions among security researchers [Neumann, 2000]. Despite the efforts of researchers in the private sector and academia to promote developer awareness and to develop tools to eliminate XSS attacks, hackers are still using them to exploit Web applications. A study by Ohmaki (2002) [Ohmaki, 2002] found that almost 80 percent of all e-commerce sites in Japan were still vulnerable to XSS. A search on Google News (http://news.google.com) for XSS advisories on newly discovered XSS vulnerabilities within the month of March 2004 alone yielded 24 reports. Among these were confirmed vulnerabilities in Microsoft Hotmail [Varghese, 2004] and Yahoo! Mail [Krishnamurthy, 2004], both of which are popular web-based email services. Figure 12.1 gives an example of an XSS.

```
$nick=$_GET['nick']; echo "Welcome, ".$nick."!"
```

Figure 12.1. Example of an XSS vulnerability.

Values for the variable $nick come from HTTP requests and are used to construct HTML output sent to the user. An example of an attacking URL would be:

```
http://www.victim.com/default.php?
nick=<script>malicious_script();</script>
```

Attackers must find ways to make victims open this URL. One strategy is to send an e-mail containing a piece of Javascript that secretly launches a hidden browser window to open this URL. Another is to embed the same Javascript inside a Web page, and when victims open the page, the script executes and secretly opens the URL. Once the PHP code shown in Figure 12.1 receives an HTTP request for the URL, it generates the compromised HTML output shown in Figure 12.2. In this strategy, the compromised output contains mali-

```
Welcome, <script>malicious_script();</script>!
```

Figure 12.2. Compromised HTML output.

cious script prepared by an attacker and delivered on behalf of a Web server. HTML output integrity is hence broken and the Javascript Same Origin Policy [Microsoft, 1997] [Netscape] is violated. Since the malicious script is delivered on behalf of the Web server, it is granted the same trust level as the Web server, which at minimum allows the script to read user cookies set by that server. This often reveals passwords or allows for session hijacking. Furthermore, if the Web server is registered in the Trusted Domain of the victim's browser, other rights (e.g., local file system access) may be granted as well.

2.2 SQL Injection

Considered more severe than XSS, SQL injection vulnerabilities occur when untrusted values are used to construct SQL commands, resulting in the execution of arbitrary SQL commands given by an attacker. Figure 12.3 shows an example. In Figure 12.3, `$HTTP_REFERER` is used to construct a SQL command.

```
$sql="INSERT INTO client_log
VALUES('$HTTP_REFERER');";
mysql_query($sql);
```

Figure 12.3. Example of a SQL injection vulnerability.

mand. The referrer field of an HTTP request is an untrusted value given by the HTTP client; an attacker can set the field to:

```
'); TRUNCATE TABLE client_log
```

This will cause the code in Figure 12.3 to construct the `$sql` variable as:

```
INSERT INTO client_log VALUES('');
TRUNCATE TABLE client_log;
```

Table "client_log" will be emptied when this SQL command is executed. This technique, which allows for the arbitrary manipulation of backend database, is responsible for the majority of successful Web application attacks.

2.3 General Script Injection

General script injection vulnerabilities are considered the most severe of the three types discussed in this chapter. They occur when untrusted data is used to call functions that manipulate system resources (e.g., in PHP: fopen(), rename(), copy(), unlink(), etc) or processes (e.g., exec()). Figure 12.4 presents a simplified version of a general script injection vulnerability. The HTTP request variable "df" is used as an argument to call fopen(), which allows arbitrary files to be opened. A subsequent code section may deliver the opened file to the HTTP client, which allows attackers to download arbitrary files. A more severe example of this vulnerability type is shown in Figure 12.5:

```
$download_file = $_POST['df'];
if($_POST['action'] == 'download')
  $fp=fopen($download_file,'rb');
```

Figure 12.4. Example of a general script injection vulnerability.

The intent for this code is to execute the validate_user.exe program in order to validate user accounts and passwords. However, since the "user" and "pass" variables are untrustworthy, the code permits the execution of arbitrary system commands. For instance, a malicious user can send an HTTP request with user="x y; NET USER foo /ADD" and pass="". As a result, the actual command becomes:

```
exec("validate_user.exe $_POST['user']
          $_POST['pass']");
```

Figure 12.5. A more severe script injection bug.

```
validate_user.exe x y; NET USER foo /ADD
```

This results in the creation of new user "foo" with logon rights.

3. Current Countermeasures

In this section we will discuss current countermeasures or approaches to ensuring Web application security. Scott and Sharp [Scott and Sharp, 2002a] [Scott and Sharp, 2002b] have asserted that Web application vulnerabilities are a) inherent in Web application programs; and b) independent of the technology in which the application in question is implemented, the security of the

Web server, and the back-end database. An intuitive solution to Web application security is to increase the awareness of secure coding practices during the code development and implementation phase. Recently, the Open Web Application Security Project (OWASP) , an open source community dedicated to promoting Web application security, released a list of the "Top Ten Most Critical Web Application Security Vulnerabilities" [OWASP, 2003]. Many organizations (including the United States Federal Trade Commission [Federal Trade Commission, 2003]) have recommended the report as a "best practice" for Web application development. VISA referenced the OWASP report in their Cardholder Information Security Program (CISP) , and now requires that all custom code be reviewed by knowledgeable reviewers before being put into production [Visa U.S.A, 2003]. These actions suggest the growth of a security auditing process—perhaps inevitable in light of the errors that even experienced programmers tend to make [Holzmann, 2002]. Arguably, vulnerabilities are less severe and easier to fix if they are discovered during or very soon after the development stage. However, the process of code auditing by reviewers who are competent enough to detect vulnerabilities is time-consuming and costly [Cowan, 2002], and there is no guarantee that such reviews are complete in that they will find every possible flaw in systems containing millions of lines of code. With today's Web applications being developed and constructed by components from sources of different trust levels (e.g., in-house, out-sourced, commercial-off-the-shelf, open-source), there is a serious need for automated mechanisms. Researchers have proposed a broad range of automated measures against XSS attacks. According to a) the development stage at which they are adopted and b) their underlying technology, these measures can be categorized into four categories—protection, testing, verification, and blended. Table 12.1 shows a comparison of each category's strengths and drawbacks.

Table 12.1. A comparison of the three different strategies for Web application security.

	Stage deployed	Immediate protection	Vulnerability identification	Runtime overhead	Side effects	Source required	Examples
Protection	Production	Yes	No	Yes	No	No	AppShield, InterDo
Testing	Production/ Development	No	Yes	No	Yes	No	WAVES, AppScan, WebInspect, ScanDo
Verification	Development	No	Yes	No	No	Yes	JIF, CQUAL
Blended	Production/ Development	Yes	Yes	Yes	No	Yes	CCured, WebSSARI

3.1 Protection Mechanisms

Installed at the deployment phase and capable of offering immediate security assurance, protection mechanisms are the most widely-adopted solution for Web application security. However, though protective technologies such as anti-virus software, network firewalls and IDSs (intrusion detection systems) offer comparatively secure protection at the host and network levels, application-level [Curphey et al., 2002] protection technologies are still in their infancies. Park and Sandhu's cookie-securing mechanism can be adopted to eliminate XSS, but it requires explicit modifications to existing Web applications. Scott and Sharp [Scott and Sharp, 2002a] [Scott and Sharp, 2002b] proposed the use of a gateway that filters invalid and malicious inputs at the application level; Sanctum's AppShield [Sanctum, 2002], Kavado's InterDo [Kavado, 2003], and a number of commercial products now offer similar strategies. Most of the leading firewall vendors are also using deep packet inspection [Dharmapurikar et al., 2003] technologies in their attempts to filter application-level traffic. According to a recent Gartner report [Stiennon, 2003], those that don't offer application-level protection will eventually "face extinction." Although application-level firewalls offer immediate assurance of Web application security, they have at least three drawbacks: a) they require careful configuration [Bobbitt, 2002], b) they blindly protect against unpredicted behavior without investigating the actual defects that compromise quality, and c) they induce runtime overhead.

3.2 Formalizing Web Application Vulnerabilities for Testing and Verification

Adopted during the development phase, software testing and verification are two established technologies for improving software quality. Though incapable of offering immediate security assurance, the two technologies can assess software quality and identify defects. To understand how they can be applied to Web applications, we have to first formally model Web application vulnerabilities. The primary objectives of information security systems are to protect confidentiality, integrity, and availability [Sandhu, 1993]. From the examples described in Section 2, it is obvious that for Web applications, compromises in integrity are the main causes of compromises in confidentiality and availability. The relationship is illustrated in Figure 12.6. When untrusted data is used to construct trusted output without sanitization, violations in data integrity occur, leading to escalations in access rights that result in availability and confidentiality compromises.

Both software testing and verification techniques can be used to identify illegal information flow-specifically, to identify violations of *noninterference*

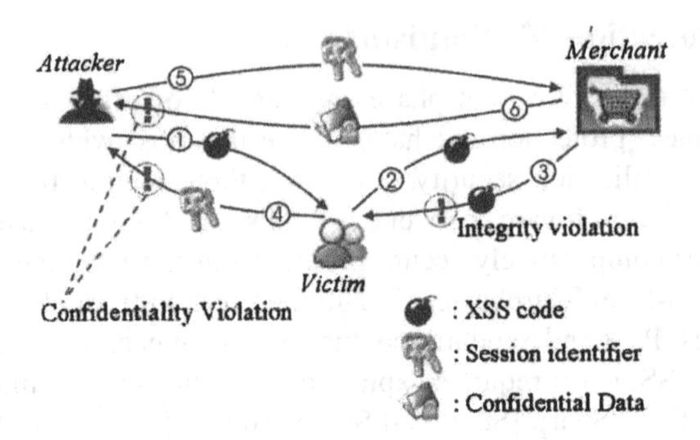

Figure 12.6. Web application vulnerabilities result from insecure information flow, as illustrated using XSS.

[Goguen and Meseguer, 1982] policies . We first make the following assumptions:

Assumption 1: All data sent by Web clients in the form of HTTP requests should be considered untrustworthy.

Assumption 2: All data local to a Web application are secure.

Assumption 3: Tainted data can be made secure with appropriate processing.

Based on these assumptions we then define the following security policies:

Policy 1: Tainted data must not be used in HTTP response construction.

Policy 2: Tainted data must not be written into local Web application storage.

Policy 3: Tainted data must not be used in system command construction.

Assumption 1 says that all data sent by Web clients (in the form of HTTP requests) should be considered untrustworthy. A majority of Web application security flaws result when this assumption is ignored or neglected. The Web uses a sessionless protocol in which each URL retrieval is considered an independent TCP session, which is established when the HTTP request is sent and terminated after a response is retrieved. Many transaction types (e.g., those that support user logins) clearly require session support. In order to keep track of sessions, Web applications require a client to include a session identifier within an HTTP request. An HTTP request consists of three major parts—the requested URL, form variables (parameters), and cookies. In practice, all three are used in different ways to store session information. Cookies are the most frequently used, followed by hidden form variables and URL requests To manage sessions, Web applications are written so that browsers include all session information following initial requests that mark the start of a session; and processing HTTP requests entails the retrieval of that information. Even though such information is transferred to the client by the Web application, it should not be considered trustworthy information when it is read back from an HTTP request. The reason is that such information is usually stored without any form of integrity protection (e.g., digital signatures), and is therefore subject to tampering. Using such information to construct HTML output without prior sanitization is considered a Policy 1 violation—the most frequent cause of XSS. Assumption 2 states that all data local to a Web application should be considered secure. This includes all files read from the file system and data retrieved from the database. According to this assumption, all locally retrieved data are considered trusted, which results in Policy 2, which states that system integrity is considered broken whenever untrustworthy data is written to local storage. Since most applications use client-supplied data to construct output, our model would be too strict without Assumption 3, which states that untrustworthy data can be made trustworthy (e.g., malicious content can be sanitized and problematic characters can be escaped). XSS vulnerabilities result in Policy 1 or Policy 2 violations. Script injection vulnerabilities such as SQL injection are generally associated with Policy 3 violations.

3.3 Software Testing for Web Application Security

For Web application security, one advantage of software testing over verification is that it considers the runtime behavior of Web applications. It is generally agreed that the massive number of runtime interactions that connect various components is what makes Web application security such a challenging task [Joshi et al., 2001] [Scott and Sharp, 2002a]. Security testing tools for Web applications are commonly referred to as *Web Security Scanners (WSS)* Commercial WSSs include Sanctum's *AppScan* [Sanctum, 2003], SPI Dynamics' *WebInspect* [SPI Dynamics, 2003], and Kavado's *ScanDo* [Kavado, 2003]. Reviews of these tools can be found in [Auronen, 2002], but to our best knowledge no literature exists on their design. Our contribution in this regard is a security assessment framework, for which we have named the *Web Application Vulnerability and Error Scanner*, or *WAVES*. We describe below WSS design challenges and solutions based on our experiences with WAVES.

3.3.1 Testing Model. All WSSs mentioned above are testing platforms posed as outsiders (i.e., as public users) to target applications. This kind of security testing is also referred to as *penetration testing* . They operate according to three constraints:

1: Neither documentation nor source code will be available for the target Web application.

2: Interactions with the target Web applications and observations of their behaviors will be done through their public interfaces, since system-level execution monitoring (e.g., software wrapping, process monitoring, and local files access) is not possible.

3: The testing process must be automated and should not require extensive human participation in test case generation.

Compared with a white-box approach (which requires source code), a black-box approach to security assessment holds many benefits in real-world applications. Consider a government entity that wishes to ensure that all Web sites within a specific network are protected against SQL injection attacks. A black-box security analysis tool can perform an assessment very quickly and produce a useful report identifying vulnerable sites. In white-box testing, analysis of source code provides critical information needed for effective test case generation [Rapps and Weyuker, 1985], whereas in black-box testing, an information-gathering approach is to reverse-engineer executable code. WSSs to date take similar approaches to identifying server-side scripts (scripts that read user input and generate output) within Web applications. These scripts constitute a Web application's *data entry points (DEPs)* . Web application interfaces that reveal DEP information include HTML forms and URLs within HTML that point to server-side scripts. In order to enumerate all DEPs of a target Web application, WSSs typically incorporate a webcrawler (also called a softbot or spider) to browse or crawl the target—an approach described in many studies involving Web site analysis (VeriWeb [Benedikt et al., 2002], Ricca and Tonella [Ricca and Tonella, 2001a] [Ricca and Tonella, 2000]) and a reverse engineering technique (Di Luca et al [Di Lucca et al., 2001] [Di Lucca et al., 2002], Ricca et al. [Ricca and Tonella, 2002] [Ricca and Tonella, 2001b] [Ricca and Tonella, 2001c]). From our experiments with WAVES, we learned that ordinary crawling mechanisms normally used for indexing purposes [Bowman et al., 1995] [Cho and Garcia-Molina, 2002] [Manber et al., 1997] [Miller and Bharat, 1998] [Sebastien] [TMIS] are unsatisfactory in terms of thoroughness. For instance, many pages within Web applications currently contain such dynamic content as Javascripts and DHTML, which cannot be handled by a webcrawler . Other applications emphasize session management, and require the use of cookies to assist navigation mechanisms. Still others require user input prior to navigation. Our tests [Huang et al., 2003] show that all traditional webcrawlers (which use static parsing and lack script interpretation abilities) tend to skip pages in Web sites that have these features. In both security assessment and fault injection, completeness is an important issue—that is, all data entry points must be correctly identified. Towards this goal, we proposed a "complete crawling" mechanism [Huang et al., 2003]—a reflection of studies on searching the hidden Web [Bergman, 2001] [Ipeirotis and Gravano, 2002] [Liddle et al., 2002] [Raghavan and Garcia-Molina, 2001] [Raghavan and Garcia-Molina, 2000].

If each DEP is defined as a program function, then each revelation is the equivalent of a function call site. We define each revelation R of a DEP as a tuple: $R = \{URL, T, Sa\}$, where URL stands for the DEP's URL, T the type of the DEP, and $Sa = \{A_1, A_2, \ldots, A_n\}$ a set of arguments (or parameters) accepted by the DEP. The type of a DEP specifies its functionality. The possible types include searching (tS), authentication (tA), account registration (tR), message posting (tM), and unknown (tU). By combining information on a DEP's URL with the names of its associated HTML forms, the names of its parameters, the names of form entities associated with those parameters, and the adjacent HTML text, WAVES [Huang et al., 2003] can make a determination of DEP type. Note that form variables are not the only sources of a DEP's input—cookies are also sources of readable input values. Therefore, the set of R's arguments $Sa = S_R \cup S_C$, where $S_R = \{P_1, P_2, \ldots, P_n\}$ is the set of parameters revealed by R, and $S_C = \{C_1, C_2, \ldots, C_n\}$ is the set of cookies contained within the page containing R.

Just as there can be multiple call sites to a program function, there may be multiple revelations of a DEP. In Google, both simple and advanced search forms are submitted to the same server-side script, with the latter submitting more parameters. We defined a DEP D as $\{dURL, dT, dSa\}$. For a set $S_D = \{R_1, R_2, \ldots, R_n\}$ of all collected revelations of the same DEP D, $dURL = R_1.URL = R_2.URL = \ldots = R_n.URL$. D's type dT = Judge_T($R_1.T$, $R_2.T$, $\ldots$, $R_n.T$), where Judge_T is a judgment function that determines a DEP's type, taking into account the types of all its revelations. D's arguments $Sa = R_1.Sa \cup R_2.Sa \cup \ldots \cup R_n.Sa$.

3.3.2 Test Case Generation. Given such a definition, a DEP can be viewed as a program function, with $dURL$ being the function name, dT the function specification, and dSa its arguments. The function output is the generated HTTP response (i.e., HTTP header, cookies, and HTML text). In this respect, testing a DEP is the same as testing a function—test cases are generated according to the function's definitions, functions are called using the test cases, and outputs are collected and analyzed.

Testing for Policy 1 violations involved using our DEP definition to generate test cases containing attack patterns, submitting them to the DEP, and studying the output for signs of the attack pattern. The appearance of an attack pattern in DEP output means that the DEP is using tainted (non-sanitized) data to construct output. The two questions guiding our test case generation were a) What is an appropriate test case size that allows for a thorough testing within an acceptable amount of time? and b) What types of test cases will/will not cause side effects?

In response to the first question, given a DEP D of $dSa = \{A_1, A_2, \ldots, A_n\}$, a naïve approach would be to generate n test cases, each with a malicious value placed in a different argument. For each test case, arguments other than the one containing malicious data would be given arbitrary values. This appears to be a reasonable approach on the surface, but it is subject to a high rate of false negatives because DEPs often execute validation procedures prior to performing their primary tasks. For example, D may use A_1 to construct output without prior sanitization, but at the beginning of its execution it will check A_2 to see if it contains a "@" character, when A_2 represents an email address. In such situations, none of our n test cases would find an error, since they would not cause D to reach its output construction phase. Instead, they would cause D to terminate early and create an error message describing A_2 as an invalid email. However, D would indeed be vulnerable. A human attacker wanting to exploit D could then supply a valid email address and learn that D uses A_1 to construct output without sanitizing it first.

To eliminate this kind of false negatives, we employed a *deep injection* mechanism in WAVES [Huang et al., 2003]. Using a *negative response extraction (NRE)* technique, the mechanism determines whether or not D uses a validation procedure. The naïve approach is used in the absence of validation. Otherwise, WAVES attempts to use its injection knowledge base to assign valid values to all arguments. Using a trial-and-error strategy, test cases are repeatedly generated and tested in an attempt to identify valid values for all arguments. If successful, then for each of the n test cases, valid values are used for arguments that do not contain malicious data. Otherwise, WAVES degrades to using the naïve approach and generates a message indicating that its test may be subject to a high false negative rate.

3.3.3 Side Effects Elimination. In [Huang et al., 2003], we acknowledged two serious deficiencies in our original WAVES design—the testing methodology had a potential side effect of causing permanent modifications (or even damage) to the state of the targeted application. For example, for every submission, a DEP D for user registration may add a new user record to a database. If D accepts ten arguments, then to test for a single malicious pattern requires generating ten test cases, with the test pattern placed at a different argument in each test case. But in practice, numerous patterns must be tested in order to provide a decent coverage. And testing for say ten malicious patterns would mean that one hundred meaningless database records would get created.

This potential side effect prevented us from performing large-scale empirical evaluations of WAVES. It should be noted that *AppScan* [Sanctum, 2003], *InterDo* [Kavado, 2003], *WebInspect* [SPI Dynamics, 2003], and similar commercial and open-source projects have the same drawback. In our subsequent efforts [Huang et al., 2004a], we added three testing modes to WAVES—heavy, relaxed, and safe modes to remedy this drawback. The heavy mode was our original mode; and side effects were simply ignored in the interest of discovering all vulnerabilities. For the two new modes, DEPs were classified according to their types into three disjoint sets S_{safe}, S_{unsafe}, and$S_{unknown}$. $\forall D \in S_{safe}$, $D.T \in${tS, tA}; $\forall D \in S_{unsafe}$, $D.T \in${tR, tM}; $\forall D \in S_{unknown}$, $D.T$=tU. In both the relaxed and safe modes, DEPs belonging to S_{unsafe} are not tested, and S_{safe} DEPs are tested using the heavy mode. In the relaxed mode, $S_{unknown}$ DEPs are tested using the malicious pattern that is most likely to reveal errors. In safe mode, these are not tested.

3.3.4 Output Observation. After submitting a test case to a DEP, its output (HTTP response) is analyzed to detect any Policy 1 violations. To avoid XSS vulnerabilities, client-submitted data containing ¡script¿ HTML tags must be processed prior to being used for output construction. Proper processing entails a) outputting errors that indicate the detection of an attack, and b) removing the tag while still processing the request, and c) encoding the ¡script¿ tag so that it is *displayed* rather than *interpreted* by the browser. To help users observe whether such sanitization steps are being taken by a DEP, we have designed test patterns so that the absence of a sanitization routine triggers the execution of a special Javascript by the browser when it renders the DEP output. An example test pattern is shown in Figure 12.7.

```
<script>alert("WAVES_TEST_1");</script>
```

Figure 12.7. An example of our test pattern for XSS.

3.3.5 Test Case Reduction. For any DEP accepting n arguments, the naïve approach requires $n \times m$ test cases for testing against m malicious patterns. To reduce the number of test cases, we modified the test patterns according to the arguments in which the patterns were placed. For example, if placed in the first argument of a DEP, the test pattern shown in Figure 12.7 will change to:

```
<script>alert("WAVES_TEST_1_ARG_1");</script>
```

This allows for the use of IE behavior to identify vulnerable arguments. Using this strategy, we placed modified versions of the same malicious pattern into all arguments of a targeted DEP. This approach requires only $1 \times m$=m case to be tested against m malicious patterns. When two or more malicious patterns appear in the output, the message box events are captured sequentially and vulnerable arguments are identified.

3.3.6 Implementation. WAVES' system architecture is shown in Figure 12.9. The webcrawlers act as interfaces between Web applications and software testing mechanisms. Without them we would not be able to apply our testing techniques to Web applications. To make the webcrawlers exhibit the same behaviors as browsers, they were equipped with IE's Document Object Model (DOM) parser and scripting engine. We chose IE's engines over others (e.g. Gecko [Mozilla] from Mozilla) because IE is the target of most attacks. User interactions with Javascript-created dialog boxes, script error pop-ups, security zone transfer warnings, cookie privacy violation warnings, dialog boxes (e.g. "Save As" and "Open With"), and authentication warnings were all logged but suppressed to ensure continuous webcrawler execution. Note that a subset of the above events is triggered by our test cases or by Web application errors. An error example is a Javascript error event produced by a scripting engine during a runtime interpretation of Javascript code. The webcrawler suppresses the dialog box that is triggered by the event and performs appropriate processing. When an event indicates an error, it logs the event and prepares corresponding entries to generate an assessment report.

When designing the webcrawler, we looked at ways that HTML pages reveal the existence of DEPs or other pages, and came up with the following list:

1. Traditional HTML anchors.
   ```
   Ex: <a href = "http://www.google.com">Google</a>
   ```
2. Framesets.
   ```
   Ex: <frame src = "http://www.google.com/
   top_frame.htm">
   ```
3. Meta refresh redirections.
   ```
   Ex: <meta http-equiv="refresh" content="0;
   URL=http://www.google.com">
   ```
4. Client-side image maps.
   ```
   Ex: <area shape="rect" href ="http://www.google.
   com">
   ```
5. Javascript variable anchors.
   ```
   Ex: document.write("\" + LangDir + "\index.htm");
   ```
6. Javascript new windows and redirections.
   ```
   Ex: window.open("\" + LangDir + "\index.htm");
   Ex: window.href = "\" + LangDir + "\index.htm";
   ```
7. Javascript event-generated executions.
   ```
   Ex: HierMenus (http://www.webreference.com)
   ```
8. Form submissions.

We established a sample site to test several commercial and academic webcrawlers, including Teleport [TMIS], WebSphinx [Miller and Bharat, 1998],

Harvest [Bowman et al., 1995], Larbin [Sebastien], Web-Glimpse [Manber et al., 1997], and Google. None were able to crawl beyond the fourth level of revelation–about one-half of the capability of the WAVES webcrawler. Revelations 5 and 6 were made possible by WAVES' ability to interpret Javascripts. Revelation 7 also refers to link-revealing Javascripts, but only following an onClick, onMouseOver, or similar user-generated event. WAVES performs an event-generation process to stimulate the behavior of active content. This allows WAVES to detect malicious components and assists in the URL discovery process. During stimulation, Javascripts located within the assigned event handlers of dynamic components are executed, possibly revealing new links. Many current Web sites incorporate DHTML menu systems to aid user navigation. These and similar Web applications contain many links that can only be identified by webcrawlers capable of handling level-7 revelations. Also note that even though the main goal of the injection knowledge manager (IKM) is to produce variable candidates so as to bypass validation procedures , the same knowledge can also be used during the crawling process . When a webcrawler encounters a form, it queries the IKM, and the data produced by the IKM is submitted by the webcrawler to the Web application for deep page discovery.

In the interest of speed, we implemented a URL hash (in memory) in order to completely eliminate disk access during the crawling process. A separate 100-record cache helped to reduce global bottlenecks at the URL hash. See also Cho and Garcia-Molina [Cho and Garcia-Molina, 2002] for a description of a similar implementation strategy. The database feeder does not insert retrieved information into the underlying database until the crawling is complete. The scheduler is responsible for managing a breadth-first crawling of targeted URLs; special care has been taken to prevent webcrawlers from inducing harmful impacts on the Web application being tested. The dispatcher directs selected target URLs to the webcrawlers and controls crawler activity. Results from crawling and injections are organized in HTML format by the report generator.

3.3.7 Experimental Results. We evaluated WAVES' DEP discovery ability by comparing its crawling (the number of pages retrieved for a target site) with other webcrawlers. From our tests [Huang et al., 2003], Teleport [TMIS] proved to be the most thorough of a group of webcrawlers that included WebSphinx [Miller and Bharat, 1998], Larbin [Sebastien], and Web-Glimpse [Manber et al., 1997]. This may be explained by Teleport's incorporation of both HTML tag parsing and regular expression-matching mechanisms, as well as its ability to statically parse Javascripts and to generate simple form submission patterns for URL discovery. On average, WAVES retrieved 28 percent more pages than Teleport when tested with a total of 14 sites [Huang et al., 2003]. We attribute the discovery of the extra pages to WAVES' script inter-

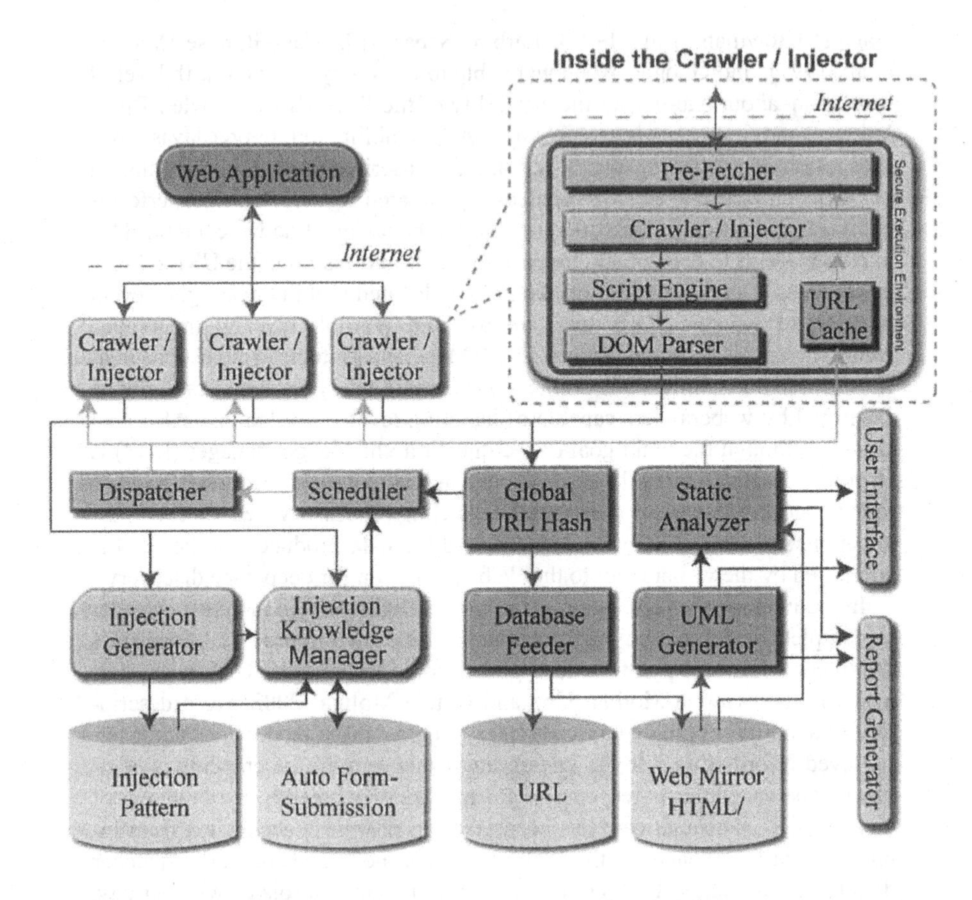

Figure 12.8. System architecture of WAVES.

pretation and automated form completion capabilities. In case study to evaluate the effectiveness of the different scanning modes we proposed, the heavy mode revealed 80 percent of all errors found by static verification [Huang et al., 2004a]. This shows that our remote, black-box testing approach provides a useful alternative to static analysis when source code and local access to the target Web application is unavailable. The 58.4 percent coverage of the relaxed mode shows that an effective non-detrimental testing is possible. The 55 strictly vulnerable sites identified during a 48-hour relaxed mode scan shows that a) our proposed mechanism for testing insecure information flow can be successfully used to detect XSS, b) non-detrimental testing still yields effective results, and c) XSS still poses a significant threat to today's Web applications. Furthermore, since tools similar to WAVES in many respects are being developed and used by hackers, we note that vulnerable websites can be easily identified by performing controlled "attacks" similar to our experiment with more malicious motivations.

3.4 Software Verification for Web Application Security

Many verification tools are discovering previously unknown vulnerabilities in legacy C programs, raising hopes that the same success can be achieved with Web applications. In Section 3.2, formalizing Web application security using three *noninterference* [Goguen and Meseguer, 1982] policies allows the use of existing verification techniques to identify Web application vulnerabilities. Sabelfeld and Myers [Sabelfeld and Myers, 2003] recently published a comprehensive survey on language-based techniques for specifying and enforcing information-flow policies. Among them, sound type systems [Volpano et al., 1996] based on the lattice model of Denning [Denning, 1976] appear most promising. Banerjee and Naumann [Banerjee and Naumann, 2002] proposed such a system for a Java-like language, and Pottier and Simonet [Pottier and Simonet, 2003] proposed one for ML. Myers [Myers, 1999] went a step further to provide an actual JIF implementation—a secure information flow verifier for the Java language. However, even though these languages can guarantee secure information flow, many consider them too strict; furthermore, they require considerable effort in terms of additional annotation in order to reduce false positives. Another problem is that most Web applications today are not developed in JIF or Java, but in script languages (e.g., PHP, ASP, Perl, and Python) [Hughes]. Using a type qualifier theory [Foster et al., 1999], Shankar et al. [Shankar et al., 2002] detected insecure information flow within legacy code with little additional annotation. Using metacompilation-based checkers [Hallem et al., 2002], Ashcraft and Engler [Ashcraft and Engler, 2002] were also able to detect insecure information flow in Linux and OpenBSD code without additional annotation. However, checkers are unsound, and both addressed only commonly found insecure information flow problems in C. To our knowledge, no comparable efforts have been made for Web applications, which involve different languages and unique information flow problems.

In contrast to compile-time techniques, run-time protection techniques are attractive because of their accuracy in detecting errors. A typical run-time approach is to instrument code with dynamic guards during the compilation phase. Cowan's Stackguard [Cowan et al., 1998] is representative of this approach; its low overhead and high accuracy has led to its inclusion in a variety of commercial software packages. Immunix Secured Linux 7+ is a commercial distribution of Linux (RedHat 7.0) that has been compiled to incorporate Stackguard instrumentation. Microsoft also includes a feature very similar to

Stackguard in its latest release of the Visual C++ .NET compiler [Microsoft, 2003].

We describe how static and runtime techniques can be used together to establish a holistic and practical approach to ensuring Web application security. We presented here our tool WebSSARI (Web application Security by Static Analysis and Runtime Inspection) [Huang et al., 2004b] [Huang et al., 2004c], which a) statically verifies existing Web application code without any additional annotation effort; and b) after verification, automatically secures potentially vulnerable sections of the code. In order to verify that Policies 1, 2 and 3 hold, WebSSARI incorporates a lattice-based static analysis algorithm derived from type systems and typestate . During the analysis, sections of code considered vulnerable are instrumented with runtime guards, thus securing Web applications in the absence of user intervention. With sufficient annotations, runtime overhead can be reduced to zero. In this section we briefly describe WebSSARI's design and our experiences learned.

3.4.1 Secure Information Flow Research. Type systems have proven useful for specifying and checking program safety properties. By means of programmer-supplied annotations, both proof-carrying codes (PCC) [Necula, 1997] and typed assembly languages (TAL) [Morrisett et al., 1999] are designed to provide safety proofs for low-level compiler-generated programs. We also used a type system to verify program security, but we targeted a high-level language, i.e., PHP, and tried to avoid additional annotations.

Many previous software security verification efforts have focused on temporal safety properties related to control flow. Schneider [Schneider, 2000] proposed formalizing security properties using *security automata* , which define the legal sequences of program actions. Walker [Walker, 2000] proposed a TAL extension, which uses security policies expressed in Schneider's automata to derive its type system. Jensen, Le Metayer and Thorn [Jensen et al., 1999] proposed using a temporal logic for specifying a program's security properties based on its control flow, and offered a model checking technique for verification. In a similar effort, Chen and Wagner [Chen and Wagner, 2002] looked for vulnerabilities in real C programs by model checking for violations of a program's *temporal safety properties*. Though their main focus was not on security, Ball and Rajamani [Ball and Rajamani, 2001] adopted a similar approach for their SLAM project and successfully applied it to Windows XP device drivers.

Type-Based Analysis. Since vulnerabilities in Web applications are primarily associated with insecure information flow, we focused our effort on ensuring proper information flow rather than control flow. The first widely accepted model for secure information flow was given by Bell and La Padula

[Bell and La Padula, 1976]. They stated two axioms: a) a subject cannot access information classified above its clearance, and b) a subject cannot write to objects classified below its clearance. Their original model only dealt with confidentiality; and Biba [Biba, 1977] is credited with adding the concept of integrity to this model.

Denning [Denning, 1976] established a lattice model for analyzing secure information flow in imperative programming languages based on a program abstraction (similar to Cousot and Cousot's [Cousot and Cousot, 1977] *abstract interpretation*) derived from an *instrumented semantics* of a language. Andrews and Reitman [Andrews and Reitman, 1980] used an axiomatic logic to reformulate Denning's model and developed a compile-time certification method using Hoare's logic. In both cases, soundness was only addressed intuitively (a more formal treatment of Denning's soundness can be found in Mizuno and Schmidt [Mizuno and Schmidt, 1992]). Orbaek [Orbaek, 1995] proposed a similar treatment, but addressed the secure information flow problem in terms of data integrity instead of confidentiality. Volpano, Smith and Irvine [Volpano et al., 1996] argued that both works proved soundness with respect to some instrumented semantics whose validity was open to question in that no means was offered for proving that the instrumented semantics correctly reflect information flow within a standard language semantics. To base directly on standard language semantics, Volpano, Smith and Irvine showed that Denning's axioms can be enforced using a type system in which program variables are associated with security classes that allow inter-variable information flow to be statically checked for correctness. Soundness was proven by showing that well-typed programs ensure confidentiality in terms of *noninterference*, a property introduced by Goguen and Meseguer [Goguen and Meseguer, 1982] for expressing information flow policies. Recently, fully functional type systems designed to ensure secure information flow have been offered for high-level, strong-typed languages such as ML [Pottier and Simonet, 2003] and Java [Myers, 1999] [Banerjee and Naumann, 2002]. Based on Foster et al.'s theory of type qualifiers [Foster et al., 1999], Shankar et al. [Shankar et al., 2002] used a constraint-based type inference engine for verifying secure information flow in C programs, and detected several format string vulnerabilities in some real C programs of which they were previously unaware.

Type-based approaches to static program analysis are attractive because they prove program correctness without unreasonable computation efforts. Their main drawback is their high false positive rates, which often makes them become impractical for real-world use. Regardless of whether security classes are assigned through manual annotations or through inference rules, in conventional type systems they are statically bound to program variables. It is important to keep in mind that the security class of a variable is a property of its state, and therefore varies at different points or call sites in a program. For example, in Myers' JIF language [Myers, 1999], each program variable is associated with a fixed security label (class). A value assumes the label of the variable in which it is stored. When a value is assigned to a variable, the value loses its original label and assumes the label of the new variable to which it is assigned. Therefore, an assignment causes a re-labeling of the security label of the assigned value. JIF ensures security by only allowing more restrictive re-labeling. However, to precisely capture information flow, values should be associated with fixed security labels, and variables should assume the labels of values they currently store—in other words, assignments should result in the re-labeling of variables rather than values. In JIF and similar type-based systems, variable labels become increasingly restrictive during computation, resulting in high false positive rates. JIF addresses this problem by giving programmers the power to *declassify* variables—that is, to explicitly relax the restrictiveness of variable labels.

Dataflow Analysis. False positives resulting from static verification of secure information flow fall into two categories. Class 1 false positives arise from the imprecise approximation of temporal variable properties. The problem described in the preceding paragraph and Doh and Shin's [Doh and Shin, 2002] *forward recovery* and *backward recovery* definitions serve as examples. In fact, most of the Denning-based systems suffer from Class 1 errors because the security class of their variables remains constant throughout program execution. Class 2 false positives result from runtime information manipulation or validation. For example, untrusted data can be sanitized before being used, with the original security class no longer applicable. This kind of false positive is more commonly associated with verifications that focus on integrity.

Class 1 errors can be reduced by making approximations of the run-time information flow more precise. Andrews and Reitman [Andrews and Reitman, 1980] first established an approach in which dataflow is semantically characterized in terms of program logic. By applying flow axioms, one can derive flow proofs that specify a program's effect on the information state. This allows the security classes of variables to change during execution, and they argued that their approach captures information flow more precisely than Denning's. Banatre, Bryce, and Le Metayer [Banatre et al., 1994] have offered a comparable approach plus a proof checking method that resembles dataflow analysis techniques associated with optimizing compilers. Joshi and Leino [Joshi and Leino, 2000] examined various logical forms for representing information flow semantics, leading to a characterization containing Hoare triples. Darvas, Hahnle, and Sands [Darvas and Hähnle, 2003] went a step further in offering characterizations in dynamic logic, which allows the use of general-purpose verifications tools (i.e., theorem provers) to analyze secure information flow within deterministic programs.

A similar approach involves flow-sensitive analysis techniques used by optimizing compilers, which have been extensively researched starting from the early works of Allen and Cocke [Allen, 1976] and followed by the works of Hecht and Ullman [Hecht and Ullman, 1973], Graham and Wegman [Graham and Wegman, 1976], Barth [Barth, 1978], and others. These methods yield more accurate runtime state predictions than the other methods mentioned above. However, flow-sensitivity comes at a price—every branch in a program's control flow doubles the verifier's search space and therefore limits its scalability. ESP , the verification tool recently developed by Das, Lerner, and Seigle [Das et al., 2002], is representative of this approach; and is based on the assumption that most program branches do not affect the information flow property that is being checked. Their contribution is distinctive because ESP allows for flow-sensitive verification that scales to large programs. They have also proposed a method called *abstract simulation* to restrict identification and simulation to relevant branch conditions. Unlike ESP , Guyer, Berger, and Lin's [Guyer et al., 2002] approach has a specific security focus. They used the flow-sensitive, context-sensitive, inter-procedural data flow analysis framework provided by their Broadway optimizing compiler to check for format string vulnerabilities of real C programs.

3.4.2 Flow-Sensitive Type-Based Analysis. A third approach emphasizes more accurate or expressive types in type systems. In their trust analysis of C programs, Shankar et al. [Shankar et al., 2002] introduced the concept of type polymorphism in their type qualifier framework, and showed how it can help reduce false positives. Others have considered extending types with state annotations. The most well known approach of this kind is Strom and Yemini's *typestate* [Strom and Yemini, 1986], which is a refinement of types. According to their definition, an object's type determines a set of allowable operations, while its typestate determines a subset allowable under specific contexts. Because it allows the flow-sensitive tracking of variable states, it serves as a technique applicable to reduce the number of Class 1 errors suffered by type-based information flow systems. Inspired by typestate , DeLine and Fahndrich [DeLine and Fahndrich, 2001] extended C types in their Vault programming language with predicates (named *type guards*) that describe legal conditions on the use of the type. In other words, types determine valid operations, while type guards determine these operations' valid times of use. In a recent project, Foster et al. [Foster, 2002] extended their original, flow-insensitive type qualifier system for C with flow-sensitive type qualifiers . Using their *Cqual* tool , they demonstrated the effectiveness of their system by discovering a number of previously unknown locking bugs in the Linux kernel.

Interestingly, the authors of ESP [Das et al., 2002] (introduced in Section 3.4.1, which tracks information flow using dataflow analysis, describe it as "merely a typestate checker for large programs." It appears that as type systems are refined with states and incorporate flow-sensitive checking, fewer differences will exist between type systems and dataflow analysis methods for verifying information flow. Our approach for reducing Class 1 errors is based primarily on typestate .

Static Checking. The goal of static *checking* is simply to find software bugs rather than to prove that one does not exist [Ashcraft and Engler, 2002]. In other words, checkers are unsound. A pioneering work was that of Bishop and Dilger [Bishop, 1996], which checked for "time-of-check-to-time-of-use" (TOCTTOU) race conditions. One recent exciting result is that of Ashcraft and Engler [Ashcraft and Engler, 2002], who used their *metacompilation* [Hallem et al., 2002] technique to find over 100 vulnerabilities in Linux and OpenBSD, over 50 of which resulted in kernel patches. The technique makes use of a flow-sensitive, context-sensitive, inter-procedural data flow checking framework that requires no additional annotations. In contrast, Flanagan et al.'s ESC/Java [Flanagan et al., 2002] (designed to check the correctness of Java programs) requires additional annotations from programmers.

Most efforts to develop checkers have resulted in publicly available tools [Cowan, 2002], including BOON by Wagner et al. [Wagner et al., 2000], RATS by Secure Software [Secure Software], FlawFinder by Wheeler [Wheeler], PScan by DeKok [DeKok], Splint by Larochelle and Evans [Larochelle and Evans, 2001] [Evans and Larochelle, 2002] , and ITS4 by Viega et al. [Viega et al., 2000]. All these unsound checkers search for specific error patterns. Splint is the only one that requires user annotations. With the exception of ESC/Java, they are all designed for use with C programs.

A Comparison. Our algorithm can be described as a sound static verification method and as a holistic method that ensures security in the absence of user intervention. Most type-based static verification methods are considered sound, provided as extensions to existing languages (e.g., Pottier and Simonet [Pottier and Simonet, 2003], Banerjee and Naumann [Banerjee and Naumann, 2002], and Myers [Myers, 1999]), and designed to support secure program development (as opposed to verifying existing code). Our work was partly inspired by the type qualifier-based verifier described in [Shankar et al., 2002] (Shankar et al.) and [Das et al., 2002] (ESP), both of which offer sound, flow-sensitive, inter-procedural data flow analysis without additional annotations. Broadway [Guyer et al., 2002] offers the same capabilities. Other checkers (e.g., MC [Ashcraft and Engler, 2002], RATS [Secure Software], and ITS4 [Viega et al., 2000]) also perform dataflow analysis without additional annotations, but their analyses are considered unsound. And as mentioned in Section 3.4.1, most of these checkers (with the exception of RATS) are targeted at C programs, while ours is targeted at PHP scripts. As its name suggests, RATS is simply a Rough Auditing Tool for Security that offers limited checks for defective PHP programming patterns; and in contrast WebSSARI offers a sound information flow analysis. Another difference is that WebSSARI ensures security by inserting runtime guards, while the other tools are limited to providing verification.

WebSSARI , MC and ITS4 are the only approaches that support *automated declassification* , defined as the process of identifying changes in a variable's security class resulting from runtime sanitization or validation. Automated declassification helps reduce the number of Class 2 false positives. MC was designed to detect sections of code that validate user-submitted integers. If the code makes both upper bound and lower bound validations on an untrusted value, it is assumed that validation has been performed; and the security class of the validated value is then changed from untrusted to trusted. This approach is based on the unsound assumption that as long as an untrusted value passes a certain kind of validation, it is actually safe. Therefore, false positives are reduced at the cost of introducing false negatives that compromise verification soundness. In the case of ITS4 , its attempt to reduce Class 2 false positives (while detecting C format string vulnerability) involves using lexical analysis to identify sanitization routines based on unsound heuristics.

When verifying information flow in Web applications, one deals with strings instead of integers, and PHP provides standard string sanitization functions. By accepting all string values processed by these functions as trusted, we first reduced a considerable number of Class 2 false positives. For cases in which custom sanitization is provided by the programmer, we proposed *type-aware qualifiers* , which resulted in a more expressive security lattice than the simple tainted-untainted lattice used by other efforts (e.g., Ashcraft and Engler [Ashcraft and Engler, 2002] and Shankar et al. [Shankar et al., 2002]), and achieved a further reduction in the number of Class 2 errors. To provide a clear representation of how our efforts compare with those of others, we have defined six criteria for classifying static analyzers: focus of scope, approach, soundness, additional annotation effort, supported language, and declassification support. A comparison based on these criteria is presented in Figure 12.9.

Runtime Protection. In many situations, it is difficult for static analysis to offer satisfactory runtime program state approximation. One strategy is to delay parts of the verification process until runtime. A good example of this practice is Perl's "tainted mode" [Wall et al., 2000], which ensures system integrity by tracking tainted data submitted by the user at runtime. In a similar manner, Myers [Myers, 1999] also leaves some JIF security class checking

	Focus	App	Snd	Anno	Lang	Dec
WebSSAR	S. I.F.	Type	Yes	Optional	PHP	Auto
CQual	S. I.F.	Type	Yes	Some	C	Manual
JIF	S. I.F.	Type	Yes	Required	Java	Manual
Vault	Gen. I.F.	Type	Yes	Required	C	Manual
ESP	Gen.	D.A.	Yes	No need	C	None
Broadway	S. I. F.	D.A.	Yes	No need	C	None
MC	S. I.F.	D.A.	No	No need	C	Auto
BOON	S.	D.A.	No	No need	C	None
ESC/Java	Gen.	D.A.	No	Required	Java	Manual
Splint	S.	L.A..	No	Required	C	Manual
ITS4	S.	L.A.	No	No need	C	Auto
MOPS	S.	Modl	Yes	No need	C	None

App—Approach Snd—Soundness
Anno—Annotation effort Lang—Supported language
Dec—Declassification support S.—Focus on security
I.F.—Focus on information flow Gen.—General verification
Type—Type system D.A.—Dataflow analysis
L.A.—Lexical analysis Modl—Model checking

Figure 12.9. A comparison among static verification tools.

operations until runtime. In dynamically typed languages such as Lisp and Scheme, a common approach is to perform runtime type checking for objects whose types have yet to be determined at compile-time. These kinds of dynamic checks are extremely expensive, resulting in the creation of such static optimization techniques as dynamic typing [Henglein, 1992] and soft typing [Wright and Cartwright, 1999] to reduce the number of runtime checks.

WebSSARI takes a similar approach—that is, by applying static analysis, it pinpoints code requiring runtime checks and inserts the checks. A similar process is found in Necula, McPeak, and Weimer's *CCured* [Necula et al., 2002]. Though not specifically focused on security, their scheme combines type inference and run-time checks to ensure type safety for existing C programs. A major difference is that our inserted guards perform sanitization tasks rather than runtime type checking—in other words, we insert sanitization routines in vulnerable sections of code that use untrusted information. If they are inserted at the proper locations, their execution time cannot be considered real overhead because the action is a necessary security check; and WebSSARI will have simply inserted lines of code omitted by a careless (or security-unaware) programmer.

3.4.3 Verification Algorithm. In PHP , which is an imperative, deterministic programming language, sets of functions may trigger violations to the three policies defined in Section 3.2. For example, exec() executes system commands, and echo() generates output. Calling exec() with tainted arguments violates Policy 3, while doing so with echo() violates Policy 1. We refer to such functions as *sensitive functions* ; and vulnerabilities will result from *tainted* (untrustworthy) data used as arguments in sensitive function calls. For each sensitive function, we intuitively derived (based on Policies 1, 2 and 3) a trust policy (expressed as a *precondition* of the function), which states the required trust level of the function's arguments. We considered all values submitted by a user as tainted (Assumption 1), and checked their propagation against a set of predefined trust policies.

Information Flow Model. To characterize data trust levels, we followed Denning's [Denning, 1976] model and made the following assumptions:

1. Each variable is associated with a security class (trust level).
2. $T = \{\tau_1, \tau_2, \ldots, \tau_n\}$ is a set of security classes.
3. T is a partially ordered set by $\leq$, which is reflexive, transitive, and anti-symmetric. For $\tau_1, \tau_2 \in T$, $\tau_1 = \tau_2 \; iff \; \tau_1 \leq \tau_2 \; and \; \tau_1 \neq \tau_2$.
4. T forms a complete lattice with a) a lower bound $\perp$ such that $\forall \tau \in T, \tau \geq \perp$ and b) an upper bound $\top$ such that $\forall \tau \in T, \tau \leq \top$.

These assumptions imply that a greatest lower bound operator and a least upper bound operator exist on T. For subset $Y \subseteq T$, let $\sqcap Y$ denote $\top$ if Y is empty and the greatest lower bound of the types in Y, otherwise; let $\sqcup Y$ denote $\perp$ if Y is empty and the least upper bound of the types in Y, otherwise.

To develop an information flow system, we need to provide a method to express the trust levels of variables. Following the lead of Foster et al. [Foster et al., 1999] and Shankar et al. [Shankar et al., 2002], we extended the existing PHP language with extra *type qualifiers*—a widely-used annotation mechanism for expressing type refinements. When used to annotate a variable, the C type qualifier const expresses the constraint that the variable can be initialized but not updated [Foster et al., 1999]. We used type qualifiers as a means for explicitly associating security classes with variables and functions. In our WebSSARI implementation, we specified preconditions for all sensitive PHP function using type qualifiers . These definitions are stored in a prelude file and loaded by WebSSARI upon startup. Another prelude file contains postconditions for functions that perform sanitization to generate trusted output from tainted input. This serves as a mechanism for automated declassifications . A third prelude file includes annotations (using type qualifiers) of all possible tainted input providers (e.g., $_GET, $_POST, $_REQUEST). Type qualifiers are also used as a means for developers to manually declassify variables. Manual declassification support is important because it allows for manual elimination of false positives, which in turn reduces the number of unnecessary runtime guards, resulting in reduce overhead.

Like Foster and Shankar we perform type inferencing (of security classes) in attempt to eliminate user annotation efforts. In conventional type-based secure information flow systems (e.g., JIF [Myers, 1999]), type inferencing is used as a means to infer the *initial* security class of a variable, and a variable is assumed to be associated with its initial security class throughout the entire program execution. As explained in Section 3.4.1, fixed variable security classes induce a large number of false positives. To develop a type system in which variable classes can change and flow-sensitive properties can be considered, we maintain our *type judgments* based on Strom and Yemini's [Strom and Yemini, 1986] typestate . A type judgment Γ is a set of mapping functions (which map variables to security classes) at a particular program point, and every program point has a unique type judgment. For each variable $x \in \mathrm{dom}(\Gamma)$, there exists a unique mapping, $\Gamma \vdash x : \tau$, and we denote the uniquely mapped type τ of x in Γ as $\Gamma(x)$. To approximate runtime typestate at compile-time, a variable's security class is viewed as a *static most restrictive* class of the variable at each point in the program text. That is, if a variable has a particular class $\tau_p$at a particular program point, then its corresponding execution time data object will have a class that is at most as restrictive as τ_p, regardless of which paths were taken to reach that point. Formally, for a set of type judgments G, we denote $\oplus G$ as the most restrictive type judgment Γ where, for all $x \in \{\mathrm{dom}(\Gamma') | \Gamma' \in G\}, \Gamma \vdash x : \sqcup Y_x$. $Y_x = \{\Gamma'(x) | \Gamma' \in G\}$ is the set of all classes of x mapped in G. When verifying a program at a particular program point, $\Gamma = \oplus G_R$, where G_R represents the set of all possible type judgments, each corresponding to a unique execution-time path that could have been taken to reach that point.

To illustrate this concept, we will use the widely-adopted tainted-untainted (T-U) lattice of security classes (e.g., by BOON [Wagner et al., 2000], Ashcraft and Engler [Ashcraft and Engler, 2002], and Shankar et al. [Shankar et al., 2002]) shown in Figure 12.10. The T-U lattice has only two elements—untainted as its lower bound and tainted as its upper bound. Assume that variable t is tainted and that variables u1 and u2 are untainted. Since exec() requires an untainted argument, for Line 2 of Figures 12.12 and 12.13 to typecheck requires that we know the static most restrictive class of X. In other words, we need to know the security class τ_{X-2} that is the most restrictive of all possible runtime classes of X at line 2, regardless of the execution path taken to get there. In line 2 of Figure 12.12, since X can be either tainted or untainted, $\tau_{X-2} = \sqcup\{\text{tainted, untainted}\}$ =tainted; line 2 therefore triggers a violation. On the other hand, line 2 of Figure 12.13 typechecks.

To preserve the static most restrictive class, rules must be defined for resolving the typestate of variable names. For the sake of simplicity, we adopted the original algorithm proposed by Strom and Yemini [Strom and Yemini, 1986]. First, we perform flow-sensitive tracking of typestate. Then at execution path merge points (e.g., the beginning of a loop or the end of a conditional statement), we define the typestate of each variable as the least upper bound of the typestates of that same variable on all merging paths. In our defined lattice (Figure 12.11), the least upper bound operator on a set selects the most restrictive class from the set. Note that while Strom and Yemini originally used typestate to represent the *static invariant* variable property, which requires applying the greatest lower bound operator, for our purpose typestate is used to represent the static most restrictive class, so we need to apply the least upper bound operator instead.

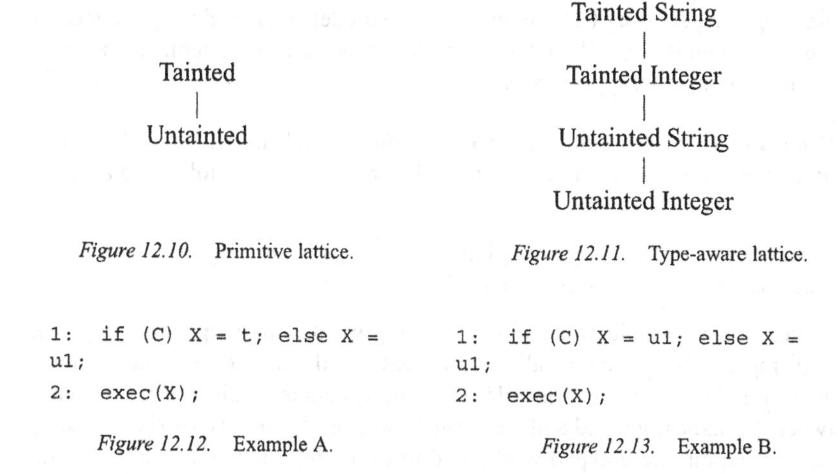

Figure 12.10. Primitive lattice.

Figure 12.11. Type-aware lattice.

Figure 12.12. Example A.

Figure 12.13. Example B.

Type-Aware Security Classes. The first version of WebSSARI implemented the verification algorithm mentioned above and made use of the T-U lattice. An initial test drive revealed a common type of false positive. Apparently many developers used type casts for sanitization purposes. An example from Obelus Helpdesk is presented in Figure 12.14. In that example, since `$_POST['index']` is tainted, $i is tainted after line 1. Line 2 therefore does not typecheck, since `echo()` requires untainted values for its argument.

```
1:  $i = (int) $_POST['index'];
2:  echo "<hidden name = mid
value='$i'>"
```

Figure 12.14. Example of a false positive resulting from a type cast.

Six of the 38 responding developers who also included copies of their intended patches for our review relied on this type of sanitization process. Since all HTTP variables are stored as strings (regardless of their actual type), using a single cast to sanitize certain variables appears to be a common practice. However, the false positive serves as evidence supporting the idea that security classes should be *type-aware*. For example, echo() can accept tainted integers without compromising system integrity (i.e., without being vulnerable to XSS). Figure 12.11 illustrates the type-aware lattice that we incorporated in our second version of WebSSARI . Until now, it has been commonly believed that annotations in type-based security systems should be provided as extensions to be checked separately from the original type system. [Foster et al., 1999] [Foster, 2002] [Shankar et al., 2002] [Flanagan et al., 2002]. In this chapter we are proposing the use of a type-aware lattice model and introducing the idea of *type-aware qualifiers*. Though still checked separately, type refinements (e.g., security classes) are type-aware.

Program Abstraction and Type Judgment. When verifying a PHP program, we first use a filter to deconstruct the program into the following abstraction:

(commands) : $c ::= c_1, c_2 | x := e | e | if\ e\ then\ c_1\ else\ c_2$
(expression) : $e ::= x | n | e_1 \sim e_2 | f(\underline{a})$

, where x is a variable, n is an integer, $\sim$ represents binary operators (e.g., +), $f(\underline{a})$ represents a function call. Commands that do not induce insecure flows are referred to as *valid* commands. The type system maintains a separation between the statically typed and the untyped worlds. To infer types (i.e., security classes) within the untyped world, and to check for command validity in the typed world, we define the following two judgment rules:
(1) Expression typing: $\Gamma \vdash e : \tau$ 2) Command validity: $\Gamma \vdash c$

Commands (which do not produce values) are distinguished from expressions (which do produce values). In these rules, Γ denotes a type judgment, which maps variables to types and also specifies the valid commands. Our type judgment rules are given below:

1. Mapping Rules:

(Initialization)

$$\frac{}{\Gamma \vdash n : \bot} \qquad \frac{x \in dom(\Gamma)}{\Gamma \vdash x : \bot}$$

(Operation)

$$\frac{\Gamma \vdash e_1 : \tau_1 \quad \Gamma \vdash e_2 : \tau_2}{\Gamma \vdash e_1 \sim e_2 : \tau_1 \sqcup \tau_2}$$

(Postcondition)

$$\frac{\Gamma \vdash f(\underline{a})}{\Gamma \vdash f(\underline{a}) : E_f(X[\underline{a}/\underline{p}])}$$

2. Checking Rule:

(Precondition)

$$\frac{f(\underline{a}), \Gamma(\underline{a}) \leq \Gamma_f(\underline{p})}{\Gamma \vdash f(\underline{a})}$$

3. Concatenation Rule:

(Concatenation)

$$\frac{\Gamma \vdash c_1 \quad \Gamma' \vdash c_2}{\Gamma, \Gamma' \vdash c_1 ; c_2}$$

4. Updating Rule:

(Assignment)

$$\frac{\Gamma \vdash x{:}\tau_1 \quad \Gamma \vdash e{:}\tau_2 \quad x := e}{\Gamma' \vdash x{:}\tau_1 \sqcup \tau_2 \quad \Gamma' \vdash x := e}$$

(Restriction)

$$\frac{\Gamma_1 \vdash c_1 \quad \Gamma_2 \vdash c_2 \quad \mathit{if}\ e\ \mathit{then}\ c_1\ \mathit{else}\ c_2}{\Gamma' = \Gamma_1 \oplus \Gamma_2 \quad \Gamma' \vdash \mathit{if}\ e\ \mathit{then}\ c_1\ \mathit{else}\ c_2}$$

The set of PHP expressions that offer tainted data and the set of sensitive functions are represented as set **I** and set **O**, respectively. To ensure secure information flow, we add the following rule that infers all expressions in **I** as tainted:

$$\frac{\forall e \in I}{\Gamma \vdash e{:}\mathrm{tainted}} \text{ (Tainted Input)}$$

We define preconditions for functions that belong to **O** as the safe sensitive function rule: $\Gamma_f(\underline{p}) = (\mathrm{untainted}, \ldots, \mathrm{untainted})$.

When verifying, we update type judgments according to command sequences and raise an error if any checking rule is violated. If we can derive a type judgment for each program point of the command sequence, we say the command sequence is secure.

Soundness. Since we always maintain the static most restrictive type judgment at every program point, a variable's type monotonically increases along the updating sequence. This is an essential property that ensures the soundness of our algorithm. However, PHP is an interpreted "scripting language" that allows for dynamic evaluation. For example, one can write "$$a" to represent a "dynamic variable," whose variable name can be determined only at runtime. To retain soundness, all dynamic variables are considered as tainted. When other kinds of dynamic evaluation exist in the target code, WebSSARI degrades itself to a checker—it still checks for potential vulnerabilities, but outputs a warning message indicating that it cannot guarantee soundness. We do, however, support pointer aliasing by implementing the original solution proposed by Strom and Yemini [Strom and Yemini, 1986]. We maintain two mappings—an environment and a store. The environment maps the names of variables involved in pointer aliasing to virtual locations, and the store maps locations to security classes. Therefore, when two pointers point to the same storage, we recognize their dereferences as a single value having a single security class. A trust level change in one pointer deference is reflected in the other.

3.4.4 System Implementation. The tool *WebSSARI* was developed to test our approach that extends an existing script language with our proposed type qualifier system. An illustration of WebSSARI's system architecture is presented in Figure 12.15. A *code walker* consists of a lexer, a parser, an AST (abstract syntax tree) maker, and a *program abstractor* . The program abstractor asks the AST maker to generate a full representation of a PHP program's AST. The AST maker uses the lexer and the parser to perform this task, handling external file inclusions along the way. By traversing the AST, the program abstractor generates a control flow graph (CFG) and a symbol table (ST). The *verification engine* moves through the CFG and references the ST to generate a) type qualifiers for variables (based on the prelude file) and b) preconditions and postconditions for functions. This routine is repeated until no new information is generated. The verification engine then moves through the control flow graph once again, this time performing typestate tracking to determine insecure information flow. It outputs insecure statements (with line numbers and the invalid arguments). For each variable involved in an insecure statement, it inserts a statement that secures the variable by treating it with a sanitization routine. The insertion is made right after the statement that caused the variable to become tainted. Sanitization routines are stored in a prelude file, and users can supply the prelude file with their own routines. Support for

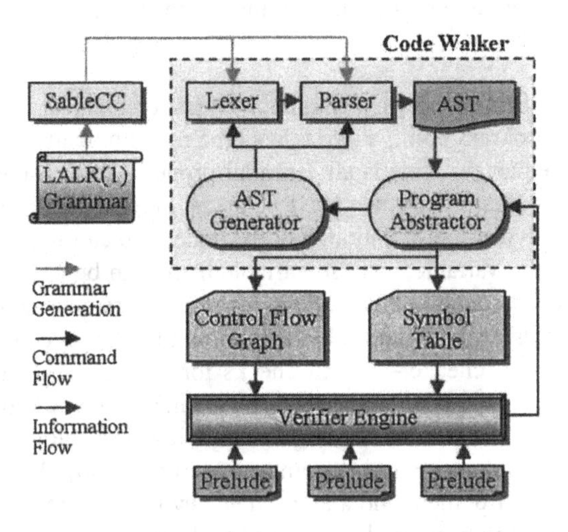

Figure 12.15. WebSSARI system architecture.

different languages is achieved by providing their corresponding code walker implementations. Since the lexers and parsers can be generated by publicly available compiler generators, providing a code walker for a language breaks down to: a) choosing a compiler generator, and providing it with the language's grammar, b) providing an AST maker, and c) providing a program abstractor . For step a), grammars for widely-used languages (e.g., C, C++, C#, and Java) are already available for widely-used compiler generators such as YACC and SableCC, and for step b), AST makers for different languages should only differ in preprocessing support (e.g., include file handling). However, since we expect considerable differences to exist in the ASTs of various languages, the major focus on providing a code walker implementation for a language is on implementing a program abstractor.

To support verification experiments using tens of thousands of PHP files, we developed a separate GUI featuring batch verification, result analysis, error logging, and report generation. Statistics can be collected based on a single source code file, files of a single project, or files of a group of projects. Vulnerable files are organized according to severity, with general script injection the most severe, SQL injection second, and XSS third. To help users investigate reported vulnerabilities, we added Watts' *PHPXREF* [Watts, 2003] to generate cross-referenced documentation of PHP source files.

In this project WebSSARI , we provided a code walker for PHP. We used Gagnon and Hendren's *SableCC* [Gagnon et al., 1998], an object-oriented compiler framework for Java. Similar to YACC and other compiler generators, SableCC accepts LALR(1) [DeRemer, 1971] grammars. No formally written grammar specifications for the PHP language exist, and no studies have been performed on whether PHP's grammar can be fully expressed in LALR(1) form. We used Mandre's [Mandre, 2003] LALR(1) PHP grammar for SableCC, which has never been thoroughly tested. The combination of SableCC and Mandre's grammar allowed us to develop a code walker for PHP; however, an initial test drive using approximately 5,000 PHP files revealed deficiencies that caused WebSSARI to reject almost 25 percent of all verified files as grammatically incorrect. With help from Mandre, we were able to reduce that rejection rate to 8 percent in a subsequent test involving 10,000 PHP files.

3.4.5 Experimental Results. SourceForge.net [Augustin et al., 2002], the world's largest open source development website, hosts over 70,000 open-source projects for more than 700,000 registered developers. PHP, currently with 7,792 registered projects, clearly outnumbers all other script languages (e.g., Perl, Python, and ASP) for Web application development. SourceForge.net classifies projects according to language, purpose, popularity, and development status (maturity). We identified a sample of 230 projects that reflected a broad variation in terms of language, purpose, popularity, and maturity. We downloaded their sources, tested them with WebSSARI, and manually inspected every report of a security violation. Where true vulnerabilities were identified, we sent email notifications to the developers. Over the five-day test period, we identified 69 projects containing real vulnerabilities; to date, 38 developers have acknowledged our findings and stated that they would provide patches. We note that in 33 of those 38 projects, the vulnerabilities had simply been overlooked, even though sanitization routines had been adopted in the majority of cases. We also found (from the developers' responses) that some of these projects had vulnerabilities that had already been identified and disclosed prior to the present project. Further inspection of their code revealed that the developers had fixed all previously published vulnerabilities, but failed to identify similar problems that were hidden throughout the code. These observations justify the need for an automated verification tool that can be used repeatedly and routinely. In all, our WebSSARI scanned 11,848 files consisting of 1,140,091 statements; and 515 files were identified as vulnerable. After four days of manual inspection, we concluded that only 361 files were indeed vulnerable-a false positive rate of 29.9 percent. The number of insecure files dropped to 494 after adding support for type-aware qualifiers, yielding a false positive rate of 26.9 percent. Type-aware qualifiers eliminated the false positive rate by 10.03 percent. Of the total 1,140,091 statements, 57,404 were associated with making calls to sensitive functions with tainted variables as arguments. WebSSARI identified 863 as insecure. After manual inspection, we concluded that 607 were actually vulnerable. Adding sanitization functions to all 57,404 statements caused 5.03 percent (57,404/1,140,091) of the 1,140,091 statements to be instrumented with dynamic guards, thus inducing overhead. After static analysis, the number of statements requiring dynamic sanitization was reduced to 863-a difference of 98.4 percent. As stated in Section 3.4.5, this instrumentation for vulnerable statements cannot be considered overhead because it simply adds code omitted by the programmer. Since only 607 statements were actually vulnerable, WebSSARI only caused 0.02 percent of all statements to be instrumented with unnecessary sanitization routines. Our experiments were conducted using a machine equipped with one Intel Pentium IV 2.0Ghz processor, 256 megabytes of RAM, and a 7,200 RPM IDE hard disk. On average, WebSSARI processed 73.85 statements per second.

4. Concluding Remarks and Future Work

Security remains a major roadblock to universal acceptance of many kinds of online transactions or services made available through the Web. This concern has been attributed to vulnerabilities of Web applications that are remotely exploitable. Many protection mechanisms are available and can offer immediate security assurance, but they induce overhead and do not address the actual software defects. On the other hand, software testing and verification are both common practices for improving software quality. In order to apply existing techniques to Web applications, Web application vulnerabilities must be formalized. In this chapter, we have formalized Web application vulnerabilities as problems involving insecure information flow, which is a conventional topic in security research. Secure information flow research was mostly motivated by confidentiality considerations; however, we have shown that Web application security require more emphasis on data integrity and trust than on confidentiality and availability. Based on our formalization, we then described how software security testing and verification could be applied to Web applications security. In software testing, researchers and engineers from the private sector have devoted a considerable amount of resources to developing WSSs , but little is known about their design challenges and their potential side effects. Another drawback is that current WSSs (including our original WAVES [Huang et al., 2003]) focus on SQL injection detection, but are deficient in XSS detection. We addressed these problems by:

1: giving a formal definition of a WSS and a list of design challenges;

2: listing test types that may induce side effects;

3: describing a test case generation process capable of producing a non-detrimental set of test cases;

4: showing how a Web application can be observed from a remote location during testing;

5: normalsizedefining three modes of remote security auditing, with a focus on potential side effects;

6: conducting an experiment using three different modes (heavy, relaxed and safe modes) and five real-world Web applications to compare differences in their coverage and induced side effects; and

7: conducting an experiment using the relaxed mode to scan random websites.

At least four assessment frameworks for Web application security (WAVES [Huang et al., 2003], AppScan [Sanctum, 2003], WebInspect [SPI Dynamics, 2003], and ScanDo [Kavado, 2003]) provide black-boxed testing capability for identifying Web application vulnerabilities. The advantage of testing over protection mechanisms is their ability to assess software quality. However, they have two disadvantages: a) they cannot provide immediate security assurance, and b) they can never guarantee soundness (they can only attempt to identify certain vulnerabilities, but cannot prove that certain vulnerabilities do not exist). By combining runtime mitigation and static verification techniques, WebSSARI demonstrates an approach that retains the advantages and eliminates the disadvantages of preceding efforts. Note that WebSSARI provides immediate protection at a much lower cost than Scott and Sharp's, since validation is restricted to potentially vulnerable sections of code. If it detects the use of untrusted data following correct treatment (e.g., sanitization), the code is left as-is. According to our experiment, WebSSARI only caused 0.02 percent of all statements to be instrumented with unnecessary sanitization routines. In contrast, Scott and Sharp [Scott and Sharp, 2002a] [Scott and Sharp, 2002b] perform unconditional global validation for every bit of user-submitted data without considering the fact that the Web application may have incorporated the same validation routine, thus resulting in unnecessary overhead. If a Web application utilizes HTTPS for traffic encryption, the associated decrypt-validate-encrypt may limit scalability. Furthermore, WebSSARI provides protection in the absence of user intervention, as compared with the user expertise required for Scott and Sharp's approach. Compared to WAVES, WebSSARI offers a sound verification of Web application code. Since verification is performed on source code, it does not require targeted Web applications to be up and running, nor is there any danger of introducing permanent state changes or loss of data. Compared to language-based approaches such as Myers [Myers, 1999], Banerjee and Naumann [Banerjee and Naumann, 2002], and Pottier and Simonet [Pottier and Simonet, 2003], our approach verifies the most commonly used language for Web application programming, and also incorporates support for extending to other languages. In other words, we provide verification for existing applications while others have proposed language frameworks for developing secure software. Their technique of typing variables to fixed classes results in a high false positive rate; while in contrast, we used typestate to perform flow-sensitive tracking that allows security classes of variables to change, resulting in more precise compile-time approximations of runtime states. Comparing flow-sensitive approaches such as Ashcraft and Engler [Ashcraft and Engler, 2002] and Shankar et al. [Shankar et al., 2002], we proposed a type-aware lattice model in contrast to their primitive tainted-untainted model. According to our experimental results, the use of this lattice model helped reduce false positives by 10.03 percent. Compared to unsound checkers [Ashcraft and Engler, 2002] [Flanagan et al., 2002] [Wagner et al., 2000] [Secure Software] [Wheeler] [DeKok] [Larochelle and Evans, 2001] [Viega et al., 2000], WebSSARI attempts to provide a sound verification framework. We note that compared with a white-box approach (which requires source code) such as WebSSARI, a black-box testing approach to security assessment (e.g., WAVES and other WSSs) still holds many benefits in real-world applications. A black-box security analysis tool can perform an assessment very quickly and produce a useful report identifying vulnerable sites. To assure high security standards, white-box testing can be used as a complement.

4.1 Future Research Directions and Open Problems

4.1.1 Protection Mechanisms—anomaly detection or misuse detection?.

The primary job of protection mechanisms such as Scott and Sharp's work [Scott and Sharp, 2002a] [Scott and Sharp, 2002b] or commercial application firewalls is to distinguish malicious traffic from normal traffic. In anomaly detection, normal traffic is defined, and those that do not comply with the definitions are considered malicious. Scott and Sharp adopted this approach for every DEP of a Web application, and required that the network administrator supply definitions describing valid parameters for the DEP. HTTP requests to a DEP that do not comply with its definition are considered malicious.

Most commercial application firewalls, on the other hand, deploy deep packet inspection [Dharmapurikar et al., 2003], which makes use of misuse detection. In misuse detection, a database of malicious patterns ("signatures") is maintained, and every HTTP request is filtered against this signature database to verify the absence of malicious data. Unfortunately, even for known attacks, neither anomaly nor misuse detection can guarantee detection. Scott and Sharp's approach asks that administrators specify valid parameters for every DEP. Though this reduces the chance of DEPs being attacked, it does not eliminate all attack possibilities. For example, the definition for an address field may be "it must be a string with length between 20 to 50 characters." A skilled attacker may still be able to exploit the DEP under this restriction—using a cleverly-crafted 20-to-50-character malicious string. On the other hand, filtering against a signature database cannot guarantee detection either. Signature-based detection has proved very successful in the anti-virus technology, because once released by its developer, a virus' executable code is fixed. However, due to the expressiveness and rich features of SQL, a same SQL injection attack can take almost an unlimited number of patterns. A detailed explanation was recently given in Maor and Shulman [Maor and Shulman, 2004]. Even if all possible attack patterns can be enumerated, real-time filtering would be impractical even with the support of advanced string filtering algorithms such as the bloom filter [Dharmapurikar et al., 2003], which is already being deployed in most application level firewalls. We note that since WebSSARI also performs signature-based filtering to sanitize untrustworthy data, it is also subject to this problem.

4.1.2 Testing—how to reduce false negatives?. WSSs available to date suffer from the high rate of false negatives due to two main reasons. Firstly, bypassing form validation procedures is difficult. Some WSSs, such as VeriWeb [Benedikt et al., 2002] and AppScan [Sanctum, 2003], adopt a profile-based solution that requires administrators to manually supply valid values for every form field. WAVES incorporates techniques associated with hidden Web crawling to address this problem. However, even with such a mechanism in place, enumerating all execution paths is difficult. For example, for many websites, a majority of DEPs will not be identified if the webcrawler does not complete a login form correctly. Even if the webcrawler is capable of recognizing a login form, the administrator must manually supply proper values (i.e., a pair of valid username and password). These suggest that manual efforts are unavoidable in order to reduce false negatives. The second reason is that current WSSs use malicious patterns to detect SQL injection vulnerabilities. They submit malicious patterns to DEPs and observe their output. A majority of these malicious patterns are designed to make the backend database of vulnerable Web applications output error messages. Such error messages are then delivered by a Web application as parts of its output and observed by the WSSs . However, many Web applications today suppress such error messages, and therefore subject current WSS testing methodology to a high rate of false negatives.

4.1.3 Verification—Web languages are hard to verify. WebSSARI incorporates a compile-time verification algorithm that statically approximates runtime state. Such approximation is harder for weakly-typed languages, and for languages that support features such as pointer aliasing, function pointers, and object-oriented programming. These features often cause the number of states of a verifier to grow exponentially, making the task of verifying larger programs nearly impossible. Popular languages used for Web application development, such as PHP and Perl, not only support all the above features, but are also scripting languages. Scripting languages are not compiled into executables but executed directly by interpreters. Therefore, they have a much looser construct and support dynamic evaluation—that is, they can generate code on the fly and have the interpreter execute them. In other words, they can programmatically interact with the underlying interpreter at runtime. All these features make it very difficult for runtime state approximation. Before the Web, complex software were seldom developed using scripting languages, and therefore not much efforts have been made to study the verification of scripting languages. However, today's Web applications are large and complex, but a majority of them are developed using scripting languages. To successfully verify these applications, techniques must be developed to handle features (such as dynamic evaluation) unique to scripting languages.

References

[Allen, 1976] Allen, F. E, Cocke, J. A Program Data Flow Analysis Procedure. Communications of the ACM, 19(3):13147, March 1976.

[Andrews and Reitman, 1980] Andrews, G. R., Reitman, R. P. An Axiomatic Approach to Information Flow in Programs. *ACM Transactions on Programming Languages and Systems*, 2(1), 56-76, 1980.

[Ashcraft and Engler, 2002] Ashcraft, K., Engler, D. Using Programmer-Written Compiler Extensions to Catch Security Holes. In *Proceedings of the 2002 IEEE Symposium on Security and Privacy*, pages 131-147, Oakland, California, 2002.

[Augustin et al., 2002] Augustin, L., Bressler, D., Smith, G. Accelerating Software Development through Collaboration. In *Proceedings of the 24th International Conference on Software Engineering*, pages 559-563, Orlando, Florida, May 19-25, 2002.

[Auronen, 2002] Auronen, L. Tool-Based Approach to Assessing Web Application Security. Helsinki University of Technology, Nov 2002.

[Ball and Rajamani, 2001] Ball, T., Rajamani, S. K., Automatically Validating Temporal Safety Properties of Interfaces. In *Proceedings of the 8th International SPIN Workshop on Model Checking of Software*, pages 103-122, volume LNCS 2057, Toronto, Canada, May 19-21, 2001. Springer-Verlag.

[Banatre et al., 1994] Banatre, J. P., Bryce, C., Le Metayer, D. Compile-time Detection of Information Flow in Sequential Programs. In *Proceedings of the Third European Symposium on Research in Computer Security*, pages 55-73, volume LNCS 875, Brighton, UK, Nov 1994. Springer-Verlag.

[Banerjee and Naumann, 2002] Banerjee, A., Naumann, D.A. Secure Information Flow and Pointer confinement in a Java-Like Language. In *Proceedings of the 15th Computer Security Foundations Workshop*, pages 239-253, Nova Scotia, Canada, 2002.

[Barth, 1978] Barth, J. M. A Practical Interprocedural Data Flow Analysis Algorithm. *Communications of the ACM*, 21(9):724-736, 1978.

[Bell and La Padula, 1976] Bell, D. E., La Padula, L. J. Secure Computer System: Unified Exposition and Multics Interpretation. Tech Rep. ESD-TR-75-306, MITRE Corporation, 1976.

[Benedikt et al., 2002] Benedikt M., Freire J., Godefroid P., VeriWeb: Automatically Testing Dynamic Web Sites. In *Proceedings of the 11th International Conference on the World Wide Web* (Honolulu, Hawaii, May 2002).

[Bergman, 2001] Bergman, M. K. The Deep Web: Surfacing Hidden Value. Deep Content Whitepaper, 2001.

[Biba, 1977] Biba, K. J. Integrity Considerations for Secure Computer Systems. Technical Report ESD-TR-76-372, USAF Electronic Systems Division, Bedford, Massachusetts, Apr 1977.

[Bishop, 1996] Bishop, M., Dilger, M. Checking for Race Conditions in File Accesses. *Computing Systems*, 9(2):131-152, Spring 1996.

[Bobbitt, 2002] Bobbitt, M. Bulletproof Web Security. *Network Security Magazine*,
TechTarget Storage Media, May 2002.
http://infosecuritymag.techtarget.com/2002/may/bulletproof.shtml

[Bowman et al., 1995] Bowman, C. M., Danzig, P., Hardy, D., Manber, U., Schwartz, M., Wessels, D. Harvest: A Scalable, Customizable Discovery and Access System. Technical Report CU-CS-732-94. , Department of Computer Science, University of Colorado, Boulder, 1995.

[CERT, 2001] CERT. CERT Advisory CA-2000-02 Malicious HTML Tags Embedded in Client Web Requests.
http://www.cgisecurity.com/articles/xss-faq.shtml

[Chen and Wagner, 2002] Chen, H., Wagner, D. MOPS: an Infrastructure for Examining Security Properties of Software. In *ACM conference on computer and communication security* (Washington, D.C., Nov 2002).

[Cho and Garcia-Molina, 2002] Cho, J., Garcia-Molina, H. Parallel Crawlers. In *Proceedings of the 11th International Conference on the World Wide Web* (Honolulu, Hawaii, May 2002), 124-135.

[Cousot and Cousot, 1977] Cousot, P., Cousot, R. Abstract Interpretation: A Unified Lattice Model for Static Analysis of Programs by Constructions or Approximation of Fixpoints. In *Conference Record of the Fourth ACM Symposium on Principles of Programming Languages*, pages 238-252, 1977.

[Cowan et al., 1998] Cowan, C., D. Maier, C. Pu, Walpole, J., Bakke, P., Beattie, S., Grier, A., Wagle, P., Zhang, Q., Hinton, H. StackGuard: Automatic adaptive detection and prevention of buffer-overflow attacks. In *Proceedings of the 7th USENIX Security Conference*, pages 63–78, San Antonio, Texas, Jan 1998.

[Cowan, 2002] Cowan, C. Software Security for Open-Source Systems. *IEEE Security and Privacy*, 1(1):38-45, 2003.

[Curphey et al., 2002] Curphey, M., Endler, D., Hau, W., Taylor, S., Smith, T., Russell, A., McKenna, G., Parke, R., McLaughlin, K., Tranter, N., Klien, A., Groves, D., By-Gad, I., Huseby, S., Eizner, M., McNamara, R. A Guide to Building Secure Web Applications. The Open Web Application Security Project, v.1.1.1, Sep 2002.

[Darvas and Hähnle, 2003] Darvas, A., Hähnle, R., Sands, D. A Theorem Proving Approach to Analysis of Secure Information Flow. In *Proceedings of the Workshop on Issues in the Theory of Security*, Warsaw, Poland, Apr 5-6, 2003.

[Das et al., 2002] Das, M., Lerner, S., Seigle, M. ESP : Path-Sensitive Program Verification in Polynomial Time. In *Proceedings of the 2002 ACM SIGPLAN Conference on Programming Language Design and Implementation*, pages 57-68, Berlin, Germany, 2002.

[DeKok] DeKok, A. PScan: A Limited Problem Scanner for C Source Files. http://www.striker.ottawa.on.ca/~aland/pscan/

[DeLine and Fahndrich, 2001] DeLine, R. Fahndrich, M. Enforcing High-Level Protocols in Low-Level Software. In *Proceedings of the ACM SIGPLAN 2001 Conference on Programming Language Design and Implementation*, pages 59-69, Snowbird, Utah, 2001.

[Denning, 1976] Denning, D. E. A Lattice Model of Secure Information Flow. *Communications of the ACM*, 19(5):236-243, 1976.

[DeRemer, 1971] DeRemer, F. Simple LR(k) Grammars. *Communications of the ACM,* 14(7):453-460, 1971.

[Dharmapurikar et al., 2003] Dharmapurikar, S., Krishnamurthy, P., Sproull, T., and Lockwood, J. Deep Packet Inspection Using Parallel Bloom Filters. In Proceedings of the 11th Symposium on High Performance Interconnects, pages 44-51, Stanford, California, 2003.

[Di Lucca et al., 2001] Di Lucca, G.A.; Di Penta, M.; Antoniol, G.; Casazza, G. An approach for reverse engineering of web-based applications. In *Proceedings of the Eighth Working Conference on Reverse Engineering* (Stuttgart, Germany, Oct 2001), 231-240.

[Di Lucca et al., 2002] Di Lucca, G.A., Fasolino, A.R., Pace, F., Tramontana, P., De Carlini, U. WARE: a tool for the reverse engineering of web applications. In *Proceedings of the Sixth European Conference on Software Maintenance and Reengineering* (Budapest, Hungary, Mar 2002), 241- 250.

[Doh and Shin, 2002] Doh, K. G., Shin, S. C. Detection of Information Leak by Data Flow Analysis. *ACM SIGPLAN Notices*, 37(8):66-71, 2002.

[Evans and Larochelle, 2002] Evans D., Larochelle, D. Improving Security Using Extensible Lightweight Static Analysis. *IEEE Software*, Jan 2002.

[Federal Trade Commission, 2003] Federal Trade Commission. Security Check: Reducing Risks to your Computer Systems. 2003. http://www.ftc.gov/bcp/conline/pubs/buspubs/security.htm

[Flanagan et al., 2002] Flanagan, C., Leino, K. R. M., Lillibridge, M., Nelson, G., Saxe, J. B., and Stata, R. Extended Static Checking for Java. In *Proceedings of the 2002 ACM SIGPLAN Conference on Programming Language Design and Implementation*, pages 234-245, volume 37(5) of ACM SIGPLAN Notices, Berlin, Germany, Jun 2002.

[Foster et al., 1999] Foster, J. S., Fähndrich, M., Aiken, A. A Theory of Type Qualifiers. In *Proceedings of the ACM SIGPLAN 1999 Conference on Programming Language Design and Implementation*, pages 192–203, volume 34(5) of ACM SIGPLAN Notices, Atlanta, Georgia, May 1-4, 1999.

[Foster, 2002] Foster, J., Terauchi, T., Aiken, A. Flow-Sensitive Type Qualifiers. In *Proceedings of the ACM SIGPLAN 2002 Conference on Programming Language Design and Implementation*, pages 1-12, Berlin, Jun 2002.

[Gagnon et al., 1998] Gagnon, E. M., Hendren, L. J. SableCC, an Object-oOiented Compiler Framework. In *Proceedings of the 1998 Conference on Technology of Object-Oriented Languages and Systems (TOOLS-98)*, pages 140-154, Santa Barbara, California, Aug 3-7, 1998.

[Goguen and Meseguer, 1982] Goguen, J. A., Meseguer, J. Security Policies and Security Models. In *Proceedings of the IEEE Symposium on Security and Privacy*, pages 11-20, Oakland, California, Apr 1982.

[Graham and Wegman, 1976] Graham, S., Wegman, M. A Fast and Usually Linear Algorithm for Global Flow Analysis. *Journal of the ACM*, 23(1):172-202, Janu 1976.

[Guyer et al., 2002] Guyer, S. Z., Berger, E. D., Lin, C. Detecting Errors with Configurable Whole-program Dataflow Analysis. Technical Report, UTCS TR-02-04, The University of Texas at Austin, 2002.

[Hallem et al., 2002] Hallem, S., Chelf, B., Xie, Y., Engler, D. A System and Language for Building System-Specific, Static Analyses. In *Proceedings of the ACM SIGPLAN 2002 Conference on Programming Language Design and Implementation*, pages 69-82, Berlin, Germany, 2002.

[Hecht and Ullman, 1973] Hecht, M. S., Ullman, J. D. Analysis of a Simple Algorithm For Global Flow Problems. In *Conference Record of the First ACM Symposium on the Principles of Programming Languages*, pages 207-217, Boston, Massachussets, 1973.

[Henglein, 1992] Henglein, F. Dynamic Typing. In *Proceedings of the Fourth European Symposium on Programming*, pages 233-253, volume LNCS 582, Rennes, France, Feb 1992. Springer-Verlag.

[Higgins et al., 2003] Higgins, M., Ahmad, D., Arnold, C. L., Dunphy, B., Prosser, M., and Weafer, V., Symantec Internet Security Threat Report—Attack Trends for Q3 and Q4 2002, Symantec, Feb 2003.

[Holzmann, 2002] Holzmann, G. J. The Logic of Bugs. In *Proceedings of the 10th ACM SIGSOFT Symposium on Foundations of Software Engineering*, pages 81-87, Charleston, South Carolina, 2002.

[Huang et al., 2003] Huang, Y. W., Huang, S. K., Lin, T. P., Tsai, C. H. Web Application Security Assessment by Fault Injection and Behavior Monitoring. In *Proceedings of the Twelfth International World Wide Web Conference*, 148-159, Budapest, Hungary, May 21-25, 2003.

[Huang et al., 2004a] Huang, Y. W., Tsai, C. H., Lee, D. T., Kuo, S. Y. Non-Detrimental Web Application Security Auditing. In *Proceedings of the Fifteenth IEEE International Symposium on Software Reliability Engineering (ISSRE2004)*, Nov 2-5, Rennes and Saint-Malo, France, 2004.

[Huang et al., 2004b] Huang, Y. W., Yu, F., Hang, C., Tsai, C. H., Lee, D. T., Kuo, S. Y. Securing Web Application Code by Static Analysis and Runtime Protection. In *Proceedings of the Thirteenth International World Wide Web Conference (WWW2004)*, pages 40-52, New York, May 17-22, 2004.

[Huang et al., 2004c] Huang, Y. W., Yu, F., Hang, C., Tsai, C. H., Lee, D. T., Kuo, S. Y. Verifying Web Applications Using Bounded Model Checking. In *Proceedings of the 2004 International Conference on Dependable Systems and Networks (DSN2004)*, pages 199-208, Florence, Italy, Jun 28-Jul 1, 2004.

[Hughes] Hughes, F. PHP: Most Popular Server-Side Web Scripting Technology. LWN.net.
http://lwn.net/Articles/1433/

[Ipeirotis and Gravano, 2002] Ipeirotis P., Gravano L., Distributed Search over the Hidden Web: Hierarchical Database Sampling and Selection. In *Proceedings of the 28th International Conference on Very Large Databases* (Hong Kong, China, Aug 2002), 394-405.

[Jensen et al., 1999] Jensen, T., Le Metayer, D., Thorn, T. Verification of Control Flow Based Security Properties. In *Proceedings of the 20th IEEE Symposium on Security and Privacy*, pages 89-103, IEEE Computer Society, New York, USA, 1999.

[Joshi and Leino, 2000] Joshi, R., Leino, K. M. A Semantic Approach to Secure Information Flow. *Science of Computer Programming*, 37(1-3):113-138, 2000.

[Joshi et al., 2001] Joshi, J., Aref, W., Ghafoor, A., Spafford, E. Security Models for Web-Based Applications. *Communications of the ACM*, 44(2), 38-44, Feb 2001.

[Kavado, 2003] Kavado, Inc. InterDo Version 3.0. Kavado Whitepaper, 2003.

[Krishnamurthy, 2004] Krishnamurthy, A. Hotmail, Yahoo in the run to rectify filter flaw. TechTree.com, March 24, 2004.
http://www.techtree.com/techtree/jsp/showstory.jsp?storyid=5038

[Larochelle and Evans, 2001] Larochelle, D., Evans, D. Statically Detecting Likely Buffer Overflow Vulnerabilites. In *Proceedings of the 10th USENIX Security Symposium*, Washington, D.C., Aug 2001.

[Liddle et al., 2002] Liddle, S., Embley, D., Scott, D., Yau, S.H., Extracting Data Behind Web Forms. In *Proceedings of the Workshop on Conceptual Modeling Approaches for e-Business* (Tampere, Finland, Oct 2002).

[Manber et al., 1997] Manber, U., Smith, M., Gopal B., WebGlimpse—Combining Browsing and Searching. In *Proceedings of the USENIX 1997 Annual Technical Conference* (Anaheim, California, Jan, 1997).

[Mandre, 2003] Mandre, I. PHP 4 Grammar for SableCC 3 Complete with Transformations. Indrek's SableCC Page, 2003.
http://www.mare.ee/indrek/sablecc/

[Maor and Shulman, 2004] Maor O., Shulman, A., SQL Injection Signatures Evasion. Imperva, Inc., Apr 2004.

[Meier et al., 2003] Meier, J.D., Mackman, A., Vasireddy, S. Dunner, M., Escamilla, R., Murukan, A. Inproving Web Application Security—Threats and Countermeasures. Microsoft Corporation, 2003.

[Microsoft, 1997] Microsoft. Scriptlet Security. Getting Started with Scriptlets, MSDN Library, 1997.
http://msdn.microsoft.com/library/default.asp?
url=/library/en-us/dnindhtm/html/instantdhtmlscriptlets.asp

[Microsoft, 2003] Microsoft. Visual C++ Compiler Options: /GS (Buffer Security Check). MSDN Library, 2003.
http://msdn.microsoft.com/library/default.asp?
url=/library/en-us/vccore/html/vclrfGSBufferSecurity.asp

[Miller and Bharat, 1998] Miller, R. C., Bharat, K. SPHINX: A Framework for Creating Personal, Site-Specific Web Crawlers. In *Proceedings of the 7th International World Wide Web Conference* (Brisbane, Australia, April 1998), 119-130.

[Mizuno and Schmidt, 1992] Mizuno, M., Schmidt, D. A. A Security Flow Control Algorithm and Its Denotational Semantics Correctness Proof. *Formal Aspects of Computing*, 4(6A):727-754, 1992.

[Morrisett et al., 1999] Morrisett, G., Walker, D., Crary, K., Glew, N. From System F to Typed Assembly Language. *ACM Transactions on Programming Languages and Systems*, 21(3):528-569, May 1999.

[Mozilla] Mozilla.org. Mozilla Layout Engine.
http://www.mozilla.org/newlayout/

[Myers, 1999] Myers, A. C. JFlow: Practical Mostly-Static Information Flow Control. In *Proceedings of the 26th ACM SIGPLAN-SIGACT Symposium on Principles of Programming Languages*, pages 228-241, San Antonio, Texas, 1999.

[Necula, 1997] Necula, G. C. Proof-Carrying Code. In *Conference Record of the 24th Annual ACM SIGPLAN-SIGACT Symposium on Principles of Programming Languages*, pages 106-119, Paris, France, Jan 1997.

[Necula et al., 2002] Necula, G. C., McPeak, S., Weimer, W. CCured: Type-Safe Retrofitting of Legacy Code. In *Proceedings of the 29th Annual ACM SIGPLAN-SIGACT Symposium on Principles of Programming Languages*, pages 128-139, Portland, Oregon, 2002.

[Netscape] Netscape. JavaScript Security in Communicator 4.x.
http://developer.netscape.com/docs/manuals/
communicator/jssec/contents.htm#1023448

[Neumann, 2000] Neumann, P. G. Risks to the Public in Computers and Related Systems. *ACM SIGSOFT Software Engineering Notes*, 25(3), p.15-23, 2000.

[Ohmaki, 2002] Ohmaki, K. Open Source Software Research Activities in AIST towards Secure Open Systems. In*Proceedings of the 7th IEEE Int'l Symp. High Assurance Systems Engineering (HASE'02)*, p.37, Tokyo, Japan, Oct 23-25, 2002.

[Orbaek, 1995] Orbaek, P. Can You Trust Your Data? In *Proceedings of the 1995 TAPSOFT/FASE Conference*, pages 575-590, volume LNCS 915, Aarhus, Denmark, May 1995. Springer-Verlag.

[OWASP, 2003] OWASP. The Ten Most Critical Web Application Security Vulnerabilities. OWASP Whitepaper, version 1.0, 2003.

[Park and Sandhu, 2002] Park, J. S., Sandhu, R. Role-Based Access Control on the Web. *ACM Transactions on Information and System Security* 4(1):37-71, 2001.

[Pottier and Simonet, 2003] Pottier, F., Simonet, V. Information Flow Inference for ML. *ACM Transactions on Programming Languages and Systems*, 25(1):117-158, 2003.

[Raghavan and Garcia-Molina, 2001] Raghavan, S., Garcia-Molina, H. Crawling the Hidden Web. In *Proceedings of the 27th VLDB Conference* (Roma, Italy, Sep 2001), 129-138.

[Raghavan and Garcia-Molina, 2000] Raghavan, S., Garcia-Molina, H. Crawling the Hidden Web. Technical Report 2000-36, Database Group, Computer Science Department, Stanford (Nov 2000).

[Rapps and Weyuker, 1985] Rapps, S., Weyuker, E. J. Selecting Software Test Data Using Data Flow Information. *IEEE Transactions on Software Engineering*, SE-11, p.367-375, 1985.

[Ricca and Tonella, 2000] Ricca, F., Tonella, P. Web Site Analysis: Structure and Evolution. In *Proceedings of the IEEE International Conference on Software Maintenance* (San Jose, California, Oct 2000), 76-86.

[Ricca and Tonella, 2001a] Ricca, F., Tonella, P. Analysis and Testing of Web Applications. In *Proceedings of the 23rd IEEE International Conference on Software Engineering* (Toronto, Ontario, Canada, May 2001), 25 –34.

[Ricca and Tonella, 2001b] Ricca, F., Tonella, P. Understanding and Restructuring Web Sites with ReWeb. *IEEE Multimedia*, 8(2), 40-51, Apr 2001.

[Ricca and Tonella, 2001c] Ricca, F., Tonella, P. Web Application Slicing. In *Proceedings of the IEEE International Conference on Software Maintenance* (Florence, Italy, Nov 2001), 148-157.

[Ricca and Tonella, 2002] Ricca, F., Tonella, P., Baxter, I. D. Restructuring Web Applications via Transformation Rules. *Information and Software Technology*, 44(13), 811-825, Oct 2002.

[Sabelfeld and Myers, 2003] Sabelfeld, A., Myers, A. C. Language-Based Information-Flow Security. *IEEE Journal on Selected Areas in Communications*, 21(1):5-19, 2003.

[Sanctum, 2003] Sanctum Inc. Web Application Security Testing – AppScan 3.5.
http://www.sanctuminc.com

[Sanctum, 2002] Sanctum Inc. AppShield 4.0 Whitepaper (2002).
http://www.sanctuminc.com

[Sandhu, 1993] Sandhu, R. S. Lattice-Based Access Control Models. *IEEE Computer*, 26(11):9-19, 1993.

[Schneider, 2000] Schneider, F. B. Enforceable Security Policies. *ACM Transactions on Information and System Security*, 3(1):30-50, Feb 2000.

[Scott and Sharp, 2002a] Scott, D., Sharp, R. Abstracting Application-Level Web Security. In *Proceedings of the 11th International Conference on the World Wide Web* (Honolulu, Hawaii, May 2002), 396-407.

[Scott and Sharp, 2002b] Scott, D., Sharp, R. Developing Secure Web Applications. *IEEE Internet Computing*, 6(6), 38-45, Nov 2,002.

[Sebastien] Sebastien@ailleret.com. Larbin – A Multi-Purpose Web Crawler.
http://larbin.sourceforge.net/index-eng.html

[Secure Software] Secure Software, Inc. RATS—Rough Auditing Tool for Security.
http://www.securesoftware.com/

[Shankar et al., 2002] Shankar, U., Talwar, K., Foster, J. S., Wagner, D. Detecting Format String Vulnerabilities with Type Qualifiers. In *Proceedings of the 10th USENIX Security Symposium*, pages 201-220, Washington DC, Aug 2002.

[SPI Dynamics, 2003] SPI Dynamics. Web Application Security Assessment. SPI Dynamics Whitepaper, 2003.

[Stiennon, 2003] Stiennon, R., Magic Quadrant for Enterprise Firewalls, 1H03. Research Note. M-20-0110, Gartner, Inc., 2003.

[Strom and Yemini, 1986] Strom, R. E., Yemini, S. A. Typestate : A Programming Language Concept for Enhancing Software Reliability. *IEEE Transactions on Software Engineering*, 12(1):157-171, Jan 1986.

[TMIS] Tennyson Maxwell Information Systems, Inc. Teleport Webspiders.
http://www.tenmax.com/teleport/home.htm

[Varghese, 2004] Varghese, S. Microsoft patches critical Hotmail hole. TheAge.com, March 24, 2004.
http://www.theage.com.au/articles/2004/03/24/
1079939690076.html

[Visa U.S.A, 2003] Visa U.S.A. Cardholder Information Security Program (CISP) Security Audit Procedures and Reporting as of 8/8/2003. Version 2.2, 2003.

[Viega et al., 2000] Viega, J., Bloch, J., Kohno, T., McGraw, G. ITS4: a static vulnerability scanner for C and C++ code. In *Proceedings of the 16th Annual Computer Security Applications Conference*, New Orleans, Louisiana, Dec 11-15, 2000.

[Volpano et al., 1996] Volpano, D., Smith, G., Irvine, C. A Sound Type System For Secure Flow Analysis. *Journal of Computer Security*, 4(3):167-187, 1996.

[Wagner et al., 2000] Wagner, D., Foster, J. S., Brewer, E. A., Aiken, A. A First Step Towards Automated Detection of Buffer Overrun Vulnerabilities. In *Proceedings of the 7th Network and Distributed System Security Symposium*, pages 3-17, San Diego, California, Feb 2000.

[Wall et al., 2000] Wall, L., Christiansen, T., Schwartz, R. L. Programming Perl. O'Reilly and Associates, 3rd edition, July 2000.

[Walker, 2000] Walker, D. A Type System for Expressive Security Policies. In *Proceedings of the 27th Symposium on Principles of Programming Languages*, pages 254-267, ACM Press, Boston, Massachusetts, Jan 2000.

[Watts, 2003] Watts, G. PHPXref: PHP Cross Referencing Documentation Generator. Sep 2003.
http://phpxref.sourceforge.net/

[Wheeler] Wheeler, D. A. FlawFinder.
http://www.dwheeler.com/flawfinder/

[Wright and Cartwright, 1999] Wright, A. K, Cartwright, R. A Practical Soft Type System for Scheme. *ACM Transactions on Programming Languages and Systems*, 19(1):87-152, Jan 1999

Chapter 13

SECURING JPEG2000 CODE-STREAMS

Robert H. Deng
Institute for Infocomm Research, Singapore 119613

Yongdong Wu
Institute for Infocomm Research, Singapore 119613

Di Ma
Institute for Infocomm Research, Singapore 119613

Abstract JPEG2000 is an emerging international standard for still image compression and is becoming the solution of choice for many digital imaging fields and applications. Part 8 of the standard, named JPSEC , is concerned with all the security aspects of JPEG2000 image code-streams, with emphasis presently on access control and authentication. An important aspect of JPEG2000 is its *"compress once, decompress many ways"* property [Taubman and Marcellin, 2000], i. e., it allows extraction of transcoded sub-images (e.g., images with various resolutions, pixel fidelities, tiles and components) from a single compressed image code-stream.

This paper presents our proposals to the JPSEC Working Group on an authentication scheme and an access control scheme for JPEG2000 image code-streams. Both schemes are fully compatible with the core part of the JPEG2000 standard. The authentication scheme possesses the so called "*sign once, verify many ways*" property. That is, it allows users to verify the authenticity and integrity of any transcoded sub-images extracted from a single code-stream protected with a single signature. The access control has the "*encrypt once, access many ways*" property. That is, it allows users access to transcoded sub-images from a single encrypted JPEG2000 code-stream.

Keywords: Access control, digital signature , JPEG2000, hash function , encryption, image compression, rooted trees.

1. Introduction

One of the well-known image compression standards is JPEG. JPEG stands for Joint Photographic Experts Group, a community of experts that was formed under the auspices of the ISO in the mid 1980's to develop a standard for still image coding. Since then, an evolution of image compression technology has taken place and a much more superior image compression standard known as JPEG2000 has been formed recently by ISO/IEC JTC/SC29/WG1 (the working group charged with the development of JPEG2000 standard). The major intention of JPEG2000 is twofold [Taubman and Marcellin, 2000, Rabbani and Joshi, 2002]. Firstly, it is designed to address most of the limitations of the original JPEG standard. Secondly, it intends to cater for the widening of application areas for JPEG technology. In addition to excellent coding performance and good error resilience , a remarkable merit of JPEG2000 is its "compress once, decompress many ways" functionality, i.e., it supports extraction of transcoded images with different resolutions, quality layers and regions-of-interest (ROIs), all from the same compressed code-stream. This functionality allows applications to manipulate or disclose only the required image data of a code-stream for any target users based on their privileges or capabilities. JPEG2000 achieves "*compress once, decompress many ways*" by encoding and organizing code-streams in a complicated but systematic way. JPEG2000 refers to all parts of the standard: Part 1 (the core) [ISO154447] is now published as an international standard, five more parts (Parts 2-6) are complete or near complete, and four new parts (Parts 8-11) are under development. In particular, Part 8 of this standard, named JPSEC, is concerned with all the security aspects of JPEG2000 image code-streams and files. At the time of this writing, the JPSEC Working Group focuses on access control and authentication mechanisms for JPEG2000 image code-streams.

This paper presents our proposals to the JPSEC Working Group on an authentication scheme and an access control scheme for JPEG2000 image code-streams. Both schemes are fully compatible with the core part of the JPEG2000 standard. The authentication scheme possesses the so called "*sign once, verify many ways*" property. That is, it allows users to verify the authenticity and integrity of any transcoded sub-images extracted from a single code-stream protected with a single signature. The access control has the so called "*encrypt once, access many ways*" property. That is, it allows user access to transcoded sub-images from a single encrypted JPEG2000 code-stream.

1.1 Related Work

Many data integrity and origin authentication techniques have been proposed [Schneier, 1996, Menezes et al., 1996]. However, these generic data authentication techniques completely ignore the internal data structure of the content under protection. A scheme using digital signature for authenticating JPEG2000 code-streams is proposed in [Grosbois et al., 2001]: it simply signs each code-block and attaches the digital signature to the end of the code-block bit stream. Hence, the scheme is neither secure nor efficient. It will generate many signatures since a code-stream may contain many code-blocks. The scheme is vulnerable to cut-and-paste attack since it only authenticates individual code-blocks, not the image code-stream as a whole. A semi-fragile JPEG image authentication scheme is presented in [Lin and Chang, 2000], which aims at authenticating images under lossy compression and other common image manipulations such as blurring and sharpening. For example, their scheme accept JPEG lossy compression on the watermarked image to a pre-determined quality factor, and reject malicious attacks. The objective of our authentication scheme is different from that of [Lin and Chang, 2000]: we aim at authenticating transcoded sub-images from a single original image code-stream. The sub-images have not only have various qualities, but also different resolutions, components and ROI.

Access control involves authorizing legitimate users with appropriate privileges to access a certain resource while denying access from illegal users [Eertino et al., 1993, Sandhu and Samarati, 1994]. Solutions for authorization fall into two categories, access control models and cryptographic techniques. An access control model mediates access to resources by checking with access control rules established in conformance with a security policy. A cryptographic method for access control manages authorization by encrypting information items such that only authorized users have the right keys to decrypt the scrambled data. A number of schemes (e.g., [Akl and Taylor, 1983, MacKinnon et al., 1985, Harn and Lin, 1990, Ohta et al., 1991, Sandhu and Samarati, 1994]) relating to access control based on cryptographic techniques have been proposed. All of these schemes assume that information items as well as users are classified into a certain type of hierarchy and there is a relationship between the encryption key assigned to a node and those assigned to its children. The related works differ mostly in the different cryptographic techniques employed for key generation. Most of them employ complex and computational expensive cryptographic operations, such as RSA [Rivest et al., 1978] or large integer modular exponentiations. Employing these schemes in access control to JPEG2000 image code-streams is not feasible since user devices (such as PDAs and cell phones) in JPEG2000 applications can be very resource constrained. The work directly relating to ours is the two access control schemes for JPEG2000 image code-streams proposed in [Grosbois et al., 2001]. The first scheme allows for a preview of low resolution image while preventing the correct display of its higher resolutions by scrambling the sign bits of the wavelet coefficients of the high resolutions code-block by code-block based on pseudo-random sequences. The second scheme provides access control to image quality layers by introducing pseudo-random noise in the higher quality layers of the image. Random seeds for generating the pseudo-random sequences are encrypted and appended to image code blocks. The two schemes in [Grosbois et al., 2001] have several serious drawbacks. Firstly, they are not flexible, providing either resolution based access control or quality layer based access control but not both at the same time. Secondly, they introduce considerable overhead to the image code-stream. Thirdly, the two schemes lead to decreased compression ratio because wavelet coefficients are randomized before compression. Finally, they are subject to several attacks [Wu and Deng, 2003].

1.2 Notations

We list below important notations used throughout the paper for ease of reference. Terminology such as precinct , resolution, resolution-increment , quality layer (or layer in short) and layer-increment will become clear in the next section. We will refer a JPEG2000 image code-stream simply as JPEG2000 code-stream or just code-stream.

n_C :the number of components in a code-stream

C_c :the cth component in a code-stream, $c = 0, 1, \ldots, n_C - 1$

n_R :the number of resolutions/resolution-increments in a code-stream

R_r :resolution-increment r in a code stream, $r = 0, 1, \ldots, n_R - 1$

n_L :the number of layers/layer-increments in a code-stream

L_l :layer-increment l in a codestream, $l = 0, 1, ..., n_L - 1$

n_P :the number of precincts in a resolution. Without loss of generality, we assume that every resolution has the same number of precincts.

P_p :the pth precinct in a resolution, $p = 0, 1, ..., n_P - 1$

$h(\cdot)$:a cryptographic one-way hash function

$x|y$:the concatenation of x and y

The rest of the paper is arranged as follows. Section 2 illustrates the basic concepts and characteristics of JPEG2000 image code-streams. Section 3 provides an overview of the schemes being studied. Sections 4 and 5 present our authentication and access control schemes, respectively. Section 6 contains our concluding remarks.

2. Overview of JPEG2000 Code-streams

In what follows, we provide a brief description of the concepts and terminology related to JPEG2000 code-streams. Our goal is to illuminate the main concepts at a sufficient level to impart an understanding of our access control scheme without dwelling into too much on the details. Those interested in the details are referred to [1-3].

2.1 Basic Concepts and Terminology

Tiles: JPEG2000 allows an image to be divided into rectangular non-overlapping regions known as tiles, which are compressed independently, as though they were entirely distinct images. Tiling reduces memory requirements during compression and decompression. For the sake of simplicity and without loss of generality, we will only consider single tile code-streams.

Components: An image is comprised of one or more components; each consists of a rectangular array of samples. For example, a RGB color image has three components with one component representing each of the red, green and blue color planes.

Resolution-increments and Resolutions: Given a component, a $(n_R - 1)$-level dyadic wavelet transform is performed. The first wavelet transform decomposes a component into four frequency subbands LL_1 (horizontally lowpass and vertically lowpass), LH_1 (horizontally lowpass and vertically highpass), HL_1 (horizontally highpass and vertically lowpass) and HH_1 (horizontally highpass and vertically highpass). The second wavelet transform further decomposes $LL1$ into another four subbands LL_2, LH_2, HL_2 and HH_2. Finally, the $(n_R - 1)$th wavelet transform decomposes LL_{n_R-2} into four subbands LL_{n_R-1}, LH_{n_R-1}, $HL_{n_R-1}$and HH_{n_R-1}. Therefore, a $(n_R - 1)$-level dyadic wavelet transform generates n_R sets of subbands, denoted as $R_0 = \{LL_{n_R-1}\}$, $R_1 = \{LH_{n_R-1}, HL_{n_R-1}, HH_{n_R-1}\}$, ..., $R_{n_R-1} = \{LH_1, HL_1, HH_1\}$. We refer to R_i as *resolution-increment* i of a code-stream.

The n_R resolution-increments above correspond to n_R image sizes or resolutions. The *resolution* 0 image is constructed from resolution-increment 0, R_0, and is a small "thumbnail" of the original image. The *resolution* 1 image is constructed from resolution-increments 0 and 1, R_0 and R_1, and is a bigger "thumbnail" of the original image. In general, the *resolution* r image is constructed from resolution-increments 0 to r, $\{R_0, R_1, \ldots, R_r\}$. Note that the resolution $n_R - 1$ image is the original image. Figure 2.1 shows $n_R = 3$ resolutions of a code-stream. The original image (resolution 2) is of size 128×128, resolution 1 is of size 64×64, and the lowest resolution 0 is of size 32×32.

Layer-increments and Layers: Following the wavelet decomposition, wavelet coefficients are quantized and each quantized subband is partitioned into small rectangular blocks, referred to as code-blocks. Each code-block is independently entropy encoded to create a compressed bit-stream which is distributed across n_L quality layers. Layers determine the quality or signal-to-noise ratio of the reconstructed image. Roughly speaking, a higher quality image needs more bits for each pixel representation than a lower quality image. Let L_0 denote the code-stream data needed to form a layer 0 image. Let L_l be the additional code-stream data to form a layer l image given L_0, L_1, ..., $L_l - 1$, $l = 1, 2, \ldots, n_L - 1$. That is, a layer l image is formed from $\{L_0, L_1, \ldots, L_{l-1}, L_l\}$.

Figure 13.1. $n_R = 3$ resolutions of an image.

Note that the image of layer $n_L - 1$ is the original image. We refer to L_l as layer-increment l, $l = 0, 1, 2, \ldots, n_L - 1$. Figure 13.1 illustrates two images of different qualities, one at 0.05 bpp (bits per pixel) and the other at 0.5 bpp.

Figure 13.2. Two images of different qualities.

<u>*Precincts*</u>: In order to provide locality for accessing certain portions (e.g., ROI) of an image, an intermediate space-frequency structure known as precinct is provided in JEPG2000. Unlike the tile and code-block partitions, the precinct partition does not affect the transformation or coding of sample data; it instead plays an important role in organizing compressed data within a code-stream.

A precinct is a collection of spatially contiguous code-blocks from all subbands at a particular resolution. In Figure 13.2a, the original image of size 512 × 512 is divided into 4 precincts of size 256 × 256 each. In Figure 13.2b, each smaller resolution includes 4 precincts with one-to-one correspondence to the 4 precincts in Figure 13.2a. For instance, the gray blocks marked A, B and C form a precinct in resolutions 2, 1, and 0, respectively, and they all correspond to the gray precinct in Figure 13.2a. In other words, the data in precincts A, B and C can be used to reconstruct the gray region in the original image.

Packets: Packets are the fundamental building blocks in a JPEG2000 code-stream. It comprises the compressed bit-stream from code blocks belonging to a specific component, resolution, layer and precinct. Figure 13.3 shows the process for generating packets. The original image is first decomposed into components. Then, the dyadic wavelet transform is applied to each component to produce the subbands corresponding to various resolution-increments. Each subband is quantized and divided into code-blocks. Certain spatially contiguous code-blocks in subbands of a resolution form a precinct. Each code-block is encoded independently into compressed bit-stream which is distributed into quality layer increments. Finally, compressed bits from the same component, resolution-increment, precinct and layer-increment are encapsulated into a packet.

2.2 Progression Orders

Progressive display allows images to be reconstructed with increasing pixel quality or resolution, as needed or desired, for different target devices. JPEG2000 supports progression in four dimensions: quality layer, resolution, spatial location and component [1-3].

1. In layer progression, image quality is improved when more layer-increments are received. For example, the image with the lowest quality is reconstructed from decoding L_0. The image with the next quality layer is obtained by decoding L_0 and L_1.

2. In resolution progression, the first few bytes, i.e., R_0, is used to form a small "thumbnail" of the image. As more resolution-increments R_r are received, $r = 1, 2, \ldots, n_R - 1$, image width and height double.

3. In spatial location progression, image is displayed in approximately raster fashion, i.e., from left to right and from top to down, or displayed by ROI.

4. Component progression controls the order in which the data corresponding to different components is decoded.

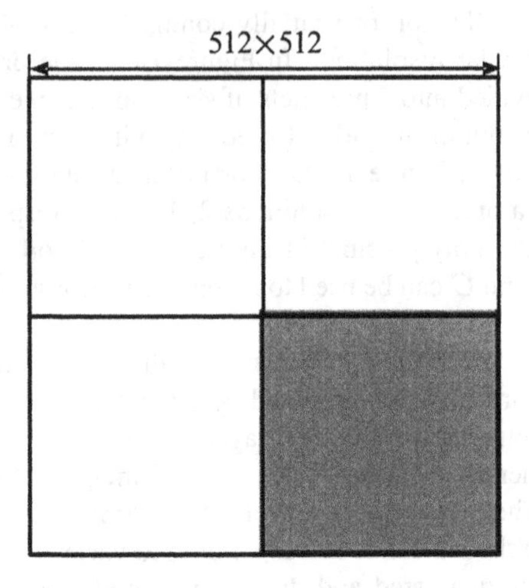

(a) Precincts of the original image

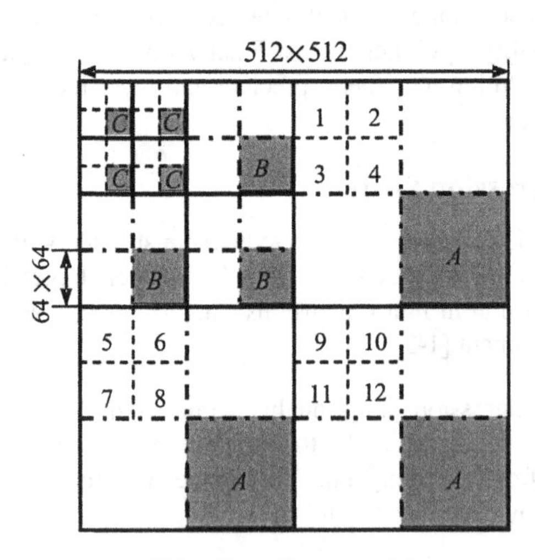

(b) Precincts of lower resolutions

Figure 13.3. Partitioning resolutions into precincts.

These dimensions of progression can be "mixed and matched" within a single compressed code-stream and this has been referred to as the "compress once, decompress many ways" property of JPEG2000 [Taubman and Marcellin, 2000]. Figure 2.2 shows a typical JPEG2000 code-stream which con-

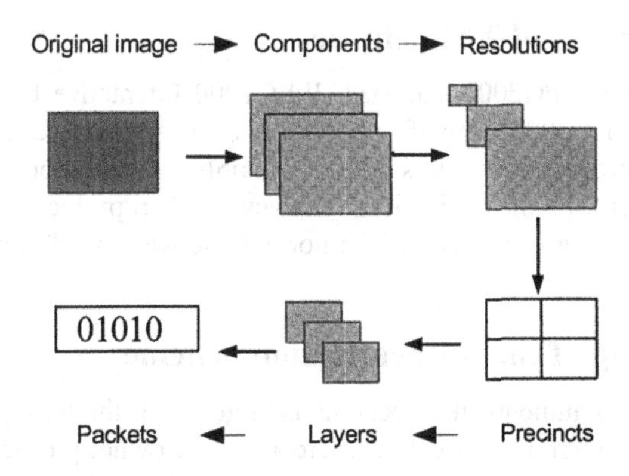

Figure 13.4. Packet generation process.

sists of a header, data packets D_1, D_2, $\ldots$, D_N arranged in a particular progression order, and an end-of-code-stream marker EOC.

Header	D_1	D_2	...	D_N	EOC

Figure 13.5. Structure of a JPEG2000 code-stream.

Figure 13.5 shows the pseudo-code for generating a code-stream of progression order layer-resolution-component-precinct. It is very important to note that a packet in a code-stream is uniquely determined by a given layer-increment l, resolution-increment r, component c, and precinct p.

for $l = 0, 1, \ldots, n_L$-1
  for $r = 0, 1, \ldots, n_R$-1
    for $c = 0, 1, \ldots, n_C$-1
      for $p = 0, 1, \ldots, n_P$-1
        packet D^{lrcp} corresponds to l, r, c, p.

Figure 13.6. Arrangement of packets in a code-stream following progression order layer-resolution-component-precinct.

3. Overview of The Schemes

Part 9 of the JEPG2000 standard, JPEG2000 Interactive Protocol (JPIP) [Prandolini et al., 2002], specifies how to respond to user requests of images with various progression orders. JPIP is mainly intended for interactive on-line user/server type of applications. However, when protected with security services, JPIP can also be adapted for non-interactive as well as off-line distributions.

3.1 Setup of the Authentication Scheme

Our proposed authentication scheme is targeted for the third party publication scenario shown in Figure 3.1, where an image owner prepares JPEG2000 code-streams for a publisher (the third party) to disseminate to users on demand. The responses from the publisher to users' requests are transcoded sub-images. In security-sensitive applications, it is highly desirable or even mandatory for users to verify the authenticity of a received response, to make sure that the sub-image is indeed originated from the owner as claimed and that the content of the sub-image has not been modified during the transmission.

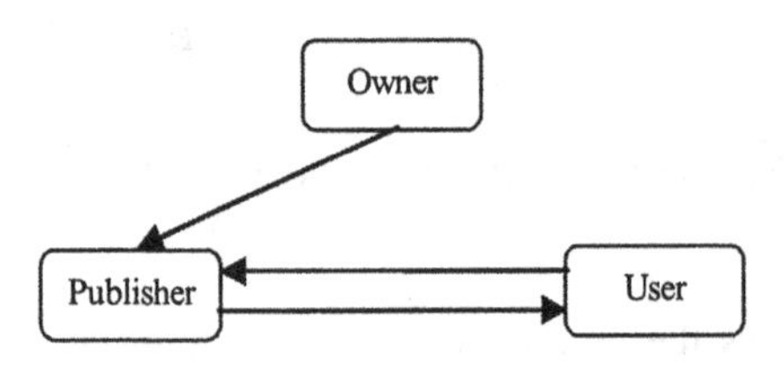

Figure 13.7. A third party publication model.

A straightforward solution is to let the publisher digitally sign each requested sub-image in real-time. This requires that the publisher be trusted by the users and does not tamper the original owner's image streams. It also requires that the private signing key be made on-line. Generally speaking, an on-line signing key is vulnerable to both external hacking and insider attacks. Another naive approach is to have the owner pre-compute signatures for all possible sub-images and forward them together with the code-stream to the publisher for distribution to users. This approach is infeasible in practice since there are too many sub-images for a code-stream. The approach is also not scalable to large number of code-streams.

Our authentication scheme presented in the next section allows for *"sign once, verify many ways"*; hence, it addresses exactly the authentication problem in the third party publication model. Using our scheme, the system in Figure 3.1 operates as follows:

1 The owner of an image code-stream prepares a digitally signed code-stream and then forwards it to the publisher.

2 A user sends the publisher a request for a sub-image of the code-stream.

3 Upon receiving the request, the publisher extracts the requested sub-image and sends the sub-image, the digital signature from the owner and a small amount of auxiliary data to the user.

4 The user verifies the authenticity of the sub-image using the digital signature and the auxiliary data. The user accepts the sub-image if the verification is successful.

Using our authentication scheme in third party publications has three important advantages: 1) the owner only needs to compute the signature once instead of pre-computing signatures for all possible sub-images, 2) the owner need not sign sub-images in real time so that much better security is achieved since the private signing key is not kept on-line and, 3) it requires less trust on the publisher since the publisher is only used for information dissemination, not for generation of digital signatures on behalf of the owner.

3.2 Setup of the Access Control Scheme

Our access control scheme aims at controlled access to JPEG2000 code-streams which are distributed following the "super-distribution" model [Mori and Kawahara, 1990]. A typical system setup is shown in Figure 3.2. High level operation of the system consists of the following three steps:

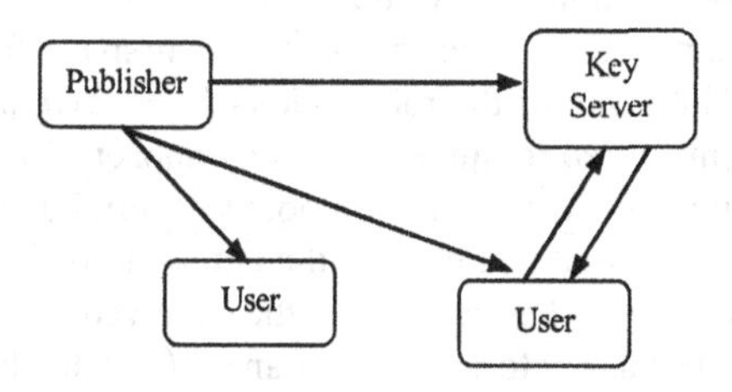

Figure 13.8. The access control system setup.

1 JPEG2000 code-streams are first encrypted by the publisher (or owner) and then distributed to users. Since the code-streams are protected by encryption, all conceivable ways of content data distribution can now be enabled, including for examples Internet, digital cable TV, satellite broadcast and CD-ROM publishing. This concept of super-distribution provides the publisher a very flexible way to use the most appropriate distribution channel.

2 The control data, i.e., keys for decrypting the content data, is forwarded securely from the publisher to an on-line key server.

3 A user who desires to access portions of a code-stream sends his/her request together with authentication information or payment data to the key server. The key server, in turn, responds with appropriate decryption keys according to user's privilege or amount of payment.

4. The Authentication Scheme

In order to preserve the "compress once, decompress many ways" property of the JPEG2000 code-streams, we require that our authentication scheme support "sign once, verify many ways". That is, the scheme allows for verification of the authenticity and integrity of any transcoded sub-images extracted from a compressed code-stream signed with a single digital signature. We achieve this design objective using the Merkle hash tree , or Merkle tree [Merkle, 1989].

4.1 The Merkle Hash Tree

To authenticate data values $n_1, n_2, \ldots, n_w$, the data source constructs the Merkle tree as depicted in Figure 4.1 assuming that $w = 4$. Each node in the tree is assigned a value. The values of the 4 leaf nodes are the message digests, $h(n_i)$, $i = 1, 2, 3, 4$, respectively, of the data values under a one-way hash function $h(\cdot)$. The value of each internal node of the tree is derived from its child nodes. For example, the value of node A is $h_a = h(h(n_1)|h(n_2))$. The data source completes the levels of the tree recursively from the leaf nodes to the root node. The value of the root node is $h_r = h(h_a|h_b)$ which is used to commit to the entire tree to authenticate any subset of the data values n_1, n_2, n_3, n_4, in conjunction with a small amount of auxiliary information. For example, a user, who is assumed to have the authentic root value h_r, requests for n_3 and requires the authentication of the received n_3. Besides n_3, the source sends the auxiliary information ha and $h(n_4)$ to the user. The user can then check the authenticity of the received n_3 as follows. The user first computes $h(n_3)$, $h_b = h(h(n_3)|h(n_4))$ and $h_r = h(h_a|h_b)$, and then checks if the latter is the same as the authentic root value h_r. Only if when this check is positive, the user accepts n_3. In general, to authenticate the data value n_i, the auxiliary information are the values of all the sibling nodes of those nodes on the path from the leaf node n_i to the root.

The concept of Merkle tree has been used for certifying answers to queries over XML documents [Devanbu et al., 2001a], for proving the presence or absence of public key certificates on revocation lists [27,28], and for certifying data published by untrusted publishers [Devanbu et al, 2001b]. However, authenticating JPEG2000 image code-streams requires more careful treatment since these streams are not as structured and are subject to various progression orders which further complicates the issue. The innovative contribution of our work is the development of a general authentication model of JPEG2000 code-streams using the Merkle tree which is compatible with JPEG2000 standard.

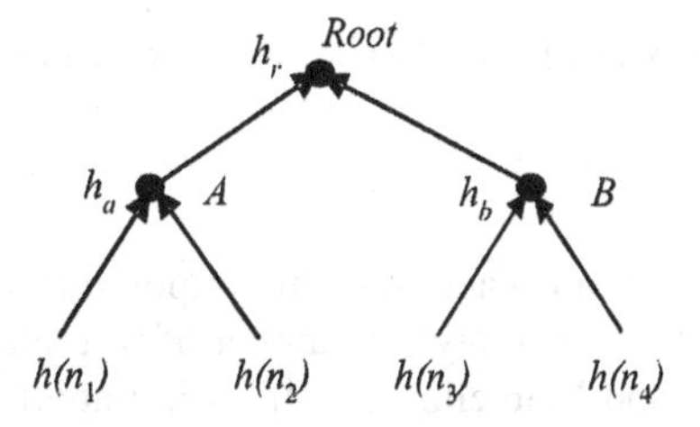

Figure 13.9. An example Merkle hash tree.

4.2 Description of the Scheme

Figures 13.3 and 13.5 show clearly a hierarchical structure of JPEG2000 code-streams. To keep our notations and drawings compact and without loss of generality, we assume that a code-stream has 1 tile and 1 component. Then a code-stream can be reviewed as a collection of precincts $\{P_p, p = 0, 1, ..., n_P - 1\}$, resolution-increments $\{R_r, r = 0, 1, ..., n_R - 1\}$, and layer-increments $\{L_l, l = 0, 1, ..., n_L - 1\}$. The Merkle tree for this code-stream is shown in Figure 4.2.

In the tree of Figure 4.2, a leaf corresponds to a code-stream packet. In a JPEG2000 code-stream, a packet is uniquely identified by a resolution, layer and precinct. Therefore, the path from the root to a leaf node in Figure 4.2 identifies a unique packet. We assign the message digest of the packet under a one-way hash function as the value of the leaf node. As an example, the path from the root to the leftmost leaf node L_0 is specified by R_0, P_0, and L_0. Hence the value of the leaf node in this path is the message digest of the packet which corresponds to resolution-increment 0, precinct 0 and layer-increment 0. Once the values of the all the leaf nodes are assigned, the values of the other nodes, including that of the root, can be computed recursively.

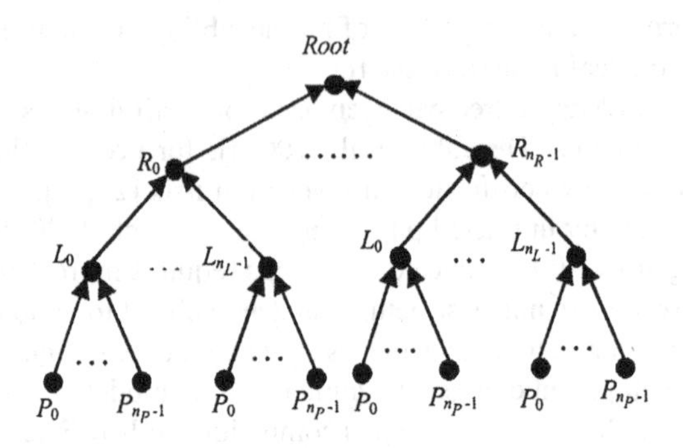

Figure 13.10. The Merkle tree for a code-stream.

A JPEG2000 code-stream header specifies important parameters of the code-stream such as size, number of layers, number of resolutions and the progression order [Taubman and Marcellin, 2000]. It is important to protect the integrity of the header in order to correctly decode the code-stream. This can be achieved by hashing the header together with the root value of the Merkle tree and let the owner sign the output of the hash function. However, to keep the presentation simple, we will not mention this explicitly in the rest of the paper.

We use an example to further illustrate the above description. Consider a code-stream with 4 resolutions (or resolution-increments), 2 layers (or layer-increments) and 2 precincts. Its Merkle hash tree is given in Figure 13.10. There are 16 leaf nodes correspond to the 16 packets in the code-stream. For example, the leftmost leaf node P_0 corresponds to the packet specified by resolution-increment 0, layer-increment 0 and precinct 0. For ease of description, we denote the 16 packets as y_0, y_1, ..., y_{15} in Figure 13.10. The owner of the code-stream assigns a value to each node in the tree according to the process described above. As an example, the leftmost P_0 node has value $h(y_1)$, the leftmost P_1 node has value $h(y_2)$, the leftmost L_0 node has a value of $h(h(y_1)|h(y_2))$. This process continues until the root value is computed. The owner generates a digital signature on the root value. The authenticated code-stream is the code-stream plus the digital signature. When a user sends a request for a transcoded sub-image with resolution 1, the owner (or a publisher as shown in Figure 3.1) sends packets $y1$, y_2, ..., y_8, the digital signature and some auxiliary data to the user. Here the auxiliary data are the values of the nodes labeled as 1 and 2 in Figure 13.10. To authenticate the received packets, the user first re-computes the root value of the hash tree based on the received packets and the auxiliary data. The user then verifies the digital signature using the owner's public key and the computed root value. The user accepts the received packets as authentic only if the verification is successful.

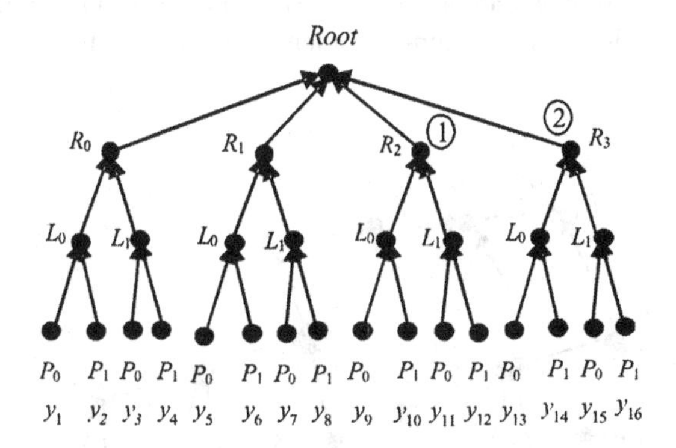

Figure 13.11. Merkle tree for an example code-stream.

A request for only one of the parameters - resolution, layer and precinct, is called a single-parameter request. They are resolution-request, layer-request and precinct-request. A request for more than one parameters is called a multiple-parameter request. In the following we consider how to optimize single-parameter requests in terms of minimizing the amount of auxiliary data. Discussion on optimizing multiple-parameter is treated in [Wu and Deng, 2003, Peng et al., 2003].

To minimize the amount of auxiliary data for resolution-request, first we note that the nodes corresponding resolution-increments should be placed at high as possible in (i. e., right below the root) the Merkle tree. Next, we remark that resolutions and resolution-increments are two different concepts (see Section 2). A resolution r sub-image is constructed from resolution-increments 0 to r, $\{R_0, R_1, \ldots, R_r\}$. A resolution-increment R_r represents the additional packets needed to construct a resolution r sub-image from a resolution $r-1$ sub-image. Therefore, a resolution-request will ask for the sets of continuous resolution-increments starting from resolution-increment 0. A similar discussion applies to layers and layer-increments. Based on the above observation, the Merkle tree in Figure 4.2 can be modified as shown in Figure 13.11, where the nodes represent resolution-increments and the layer-increments, respectively, are chained together to reflect their incremental relationships. As in Figure 4.2, each leaf node here is assigned the message digest of a unique packet. However, there are multiple nodes of the same type on a path from the root to a leaf node. For example, nodes on the path from the root to the second left P_0 node are R_0, L_0, L_1, P_0. There are two nodes of type L. Hence the method of mapping packets to leaf nodes as used for Figure 4.2 need to be modified. The modification is simple, we just ingore the nodes of the same type except the one closest to the leaf node. In the example above, we ingore L_0, so the nodes on the pruned path are R_0 L_1, P_0 and the leaf node corresponds to the packet specified by resolution-increment 0, layer-increment 1 and precinct 0.

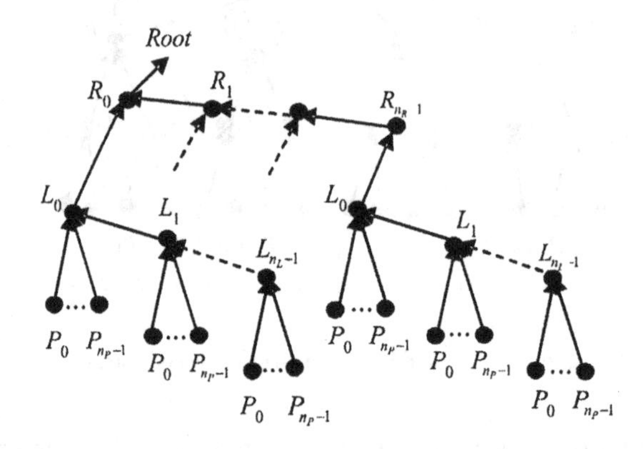

Figure 13.12. The optimized Merkle tree.

Consider again the same code-stream with 4 resolutions (or resolution-increments), 2 layers (or layer-increments) and 2 precincts. Its optimized Merkle hash tree is shown in Figure 13. When a user requests for the sub-image with resolution 1, the owner sends packets y_1, y_2, ..., y_8, the digital signature and the value of the node labelled with 1 as the auxiliary data. Note that to authenticate the same sub-image, the tree in Figure 13.10 needs two message digests while the tree here requires only one message digest. This code-stream is just a toy example. The JPEG2000 standard allows a code-stream to support up to 33 resolution-increments and 65535 layer-increments. As the numbers of resolution and layer increments increase, the amount of reduction on the auxiliary data becomes significant using the optimized Merkle tree.

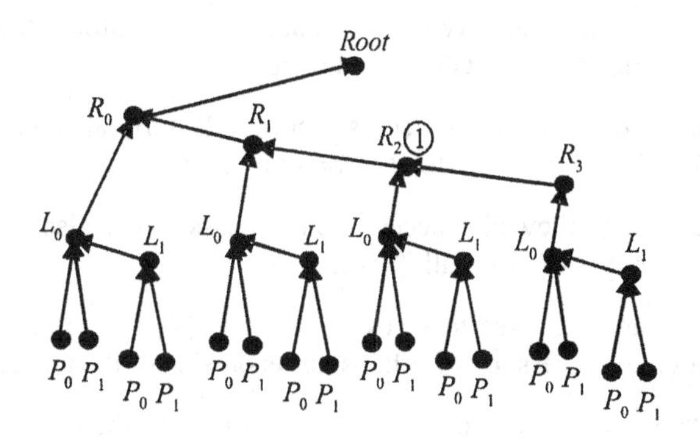

Figure 13.13. An example optimized Merkle tree.

5. The Access Control Schemes

In this section, we first define the security classes in a JPEG2000 code-stream. We then describe our access control scheme.

5.1 Security Classes and Access Control Rules

Consider the situation where users and data can be classified into a hierarchy of security classes [Sandhu, 1993]. If a security class A is the predecessor of another security class B, we say that A strictly dominates B and denote this relation as $A > B$. Similarly, we say that A dominates B, denoted as $A \geq B$, if either $A > B$ or $A = B$. We say that A and B are comparable if $A \geq B$ or $B \geq A$; otherwise A and B are incomparable. From the progression properties of code-streams presented in Section 2.2, we define the following security classes related to a JPEG2000 code-stream:

1 The security classes of resolution-increments, $\{R_r, r = 0, 1, \ldots, n_R - 1\}$, is a total ordering [Sandhu, 1993], with

$$R_{n_R-1} > R_{n_R-2} > \ldots > R_1 > R_0 \tag{13.1}$$

2 The security classes of layer-increments, $\{L_l, l = 0, 1, \ldots, n_L - 1\}$, is a total ordering, with

$$L_{n_L-1} > L_{n_L-2} > ... > L_1 > L_0 \tag{13.2}$$

3 The security classes of the precincts, $\{P_p, p = 0, 1, ..., n_P - 1\}$, are isolated classes [Sandhu, 1993]. That is,

$$P_p \text{ and } P_{p'} \text{ are incomparable for all } p \neq p'. \tag{13.3}$$

Based on the above security classes, our aim is to enforce the following access control rules for a JPEG2000 code-stream:

1 A user who is allowed to access security class R_r also have access to $R_{r'}$ for all $r' < r$ but not to $R_{r''}$ for all $r'' > r$.

2 A user who is allowed to access security class L_l can also access $L_{l'}$ for all $l' < l$ but not $L_{l''}$ for all $l'' > l$.

3 A user who is allowed to access a subset of $\{P_p, p = 0, 1, ..., n_P - 1\}$ can not have access to any other subsets outside of the granted subset.

4 Any "mix and match" of the above regardless of the progression order of the code-stream.

Care must be taken in designing access control schemes for JPEG2000 code-streams in order to avoid collusion attacks [Wu and Deng, 2003]. One approach to realize secure access control which meets our access control rules above is, using the method given in [Sandhu, 1993], to form a combined hierarchy of security classes from the isolated precinct security classes, the total ordered resolution-increment security classes and layer-increment security classes. Unfortunately, the resulting hierarchy of security classes is a Directed Acyclic Graph (DAG), not a rooted tree. There are cryptographic solutions available in the literature for key generation and implementing access controls in DAGs (see for examples [7-11, 18]). All of them, however, are based on public key cryptosystems and are extremely complex and computationally expensive for our applications.

5.2 Description of the Scheme

Sandhu [Sandhu, 1998] introduces a cryptographic implementation for access control in a situation where users and information items are classified into a rooted tree of security classes, with the most privileged security class at the root. The idea is that keys for security classes are generated from their parent class using a parameterized family of one-way functions. In the following, we seek to adapt Sandhu's approach to arrive at a flexible, efficient and secure access control scheme for JPEG2000 code-streams.

In the Sandhu tree, encryption keys associated with a tree of security classes are generated as follows: 1) for the security class at the root assign an arbitrary key; 2) if a security class Y is an immediate child of X in the tree, let $k_Y = h(k_X|name(Y))$, where k_X and k_Y are the keys associated with X and Y, respectively, and $name(Y)$ is the name of Y. The keys k_X and k_Y are used to encrypt information items of security classes X and Y, respectively. A user at a security level, say X, needs to know k_X. Since one-way hash function is public, keys for security classes dominated by X can be generated from

k_X as needed. However, it is computationally infeasible to compute keys for any predecessors or any siblings of X due to the one-way nature of the hash function.

We show the construction and application of Sandhu trees with a simple example. Consider the security classes A, B_0, B_1, C_0, C_1, C_2, and C_3, where $A > B_0$ and B_1; $B_0 > C_0$ and C_1; and $B_1 > C_2$ and C_3. The Sandhu tree for this security class hierarchy is given in Figure 13.14. We assign a random key k_A to the root. The keys for B_0 and B_1 are $k_{B_0} = h(k_A|name(B_0))$ and $k_{B_1} = h(k_A|name(B_1))$, respectively. The keys for C_0, C_1, C_2, and C_3 are $k_{C_0} = h(k_{B_0}|name(C_0))$, $k_{C_1} = h(k_{B_0}|name(C_1))$, $k_{C_2} = h(k_{B_1}|name(C_2))$ and $k_{C_3} = h(k_{B_1}|name(C_3))$, respectively. A user with security clearance B_1 is given k_{B_1}. He can easily compute keys for C_2 and C_3, but not for any other nodes in the tree.

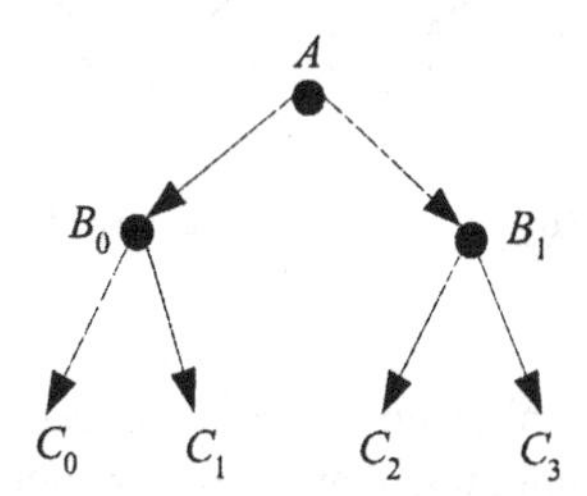

Figure 13.14. An example Sandhu tree.

Our challenge here is to bring the advantages of Sandhu tree to the access control of JPEG2000 code-streams. The key in meeting this challenge is to construct a rooted tree hierarchy of security classes for JPEG2000 code-steams. It turns out that this can be done in a systematic way. The trick is to specify a preferred progression order when constructing the overall hierarchy for a JPEG2000 code-stream.

We use an example to illustrate our idea. Assume that the preferred progression order is resolution-layer-precinct, the rooted tree hierarchy for a JPEG2000 code-stream is shown in Figure 13.15 (the dependence of the tree on the progression order will be made clearer in Section 5.3). Observe how the tree is constructed based on the preferred progression order resolution-layer-precinct: the resolution-increments form the trunk of the tree, the layer-increments form the branches and finally the precincts form the leaves. Also observe that the tree preserves the hierarchies of individual security classes, e.g., the trunk formed from the resolution-increments is still a total ordering.

We remark that there are a number of subtle differences between our tree and a Sandhu tree. First, a given class, say P_0, is assigned to multiple nodes

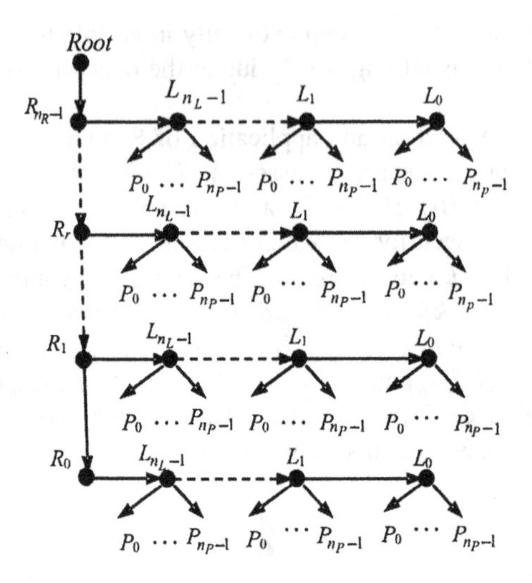

Figure 13.15. Rooted tree for key generation for access control.

in our tree while this is not allowed in a Sandhu tree. Second, keys associated with non-leaf nodes in a Sandhu tree are used for encrypting information items associated with the nodes while in our tree only keys associated with the leaf-nodes will be used to encrypt packets in the code-stream.

Packet key generation and encryption: Key generation in our rooted tree follows the same approach as that in the Sandhu tree:

1 Given a code-stream, generate a random master key K.

2 Generate keys for the resolution nodes iteratively from the hash chain

$$k^r = h(k^{r+1}), \tag{13.4}$$

for $r = n_R - 2, n_R - 3, \ldots, 1, 0$, where $k^{n_R-1} = h(K|\text{"}R\text{"})$ and where "R" denotes the ASCII code of the letter R.

3 For a given r, generate keys for the layer nodes iteratively from the hash chain

$$k^{rl} = h(k^{r(l+1)}), \tag{13.5}$$

for $r = n_R - 1, n_R - 2, \ldots, 1, 0$, $l = n_L - 2, n_L - 3, \ldots, 1, 0$, where $k^{r(n_L-1)} = h(k^r|\text{"}L\text{"})$ and where "L" denotes the ASCII code of the letter L.

4 For a given r and l, generate keys for the precinct nodes from

$$k^{rlp} = h(k^{rl}|\text{“}P\text{”}|p), \tag{13.6}$$

for $r = n_R - 1, n_R - 2, \ldots, 1, 0$, $l = n_L - 1, n_L - 2, \ldots, 1, 0$ and $p = 0, 1, \ldots, n_P - 2, n_P - 1$, where “$P$” denotes the ASCII code of the letter P.

5 Encrypt the packet D^{rlp} using the key k^{rlp} under a symmetric key algorithm for $r = 0, 1, ..., n_R - 1$, $l = 0, 1, ..., n_L - 1$ and $p = 0, 1, ..., n_P - 1$.

Access encrypted code-stream: Refer to the setup of the access control scheme in Figure 3.2, there are three cases to consider here depending on the user access requirements.

Case 1: A user requests access to the image of resolution r' (i.e., the image of resolution r' with the highest quality layer and all the precincts). The key server replies with $k^{r'}$. To obtain the packets corresponding to the requested image, the user proceeds as follows:

1 Compute, from $k^{r'}$, keys $k^{r'-1}$, $k^{r'-2}$, ..., k^1, k^0, iteratively using (13.4).

2 Compute, from the keys obtained in step 1, keys k^{rl}, iteratively using (13.5), for $r = r', \ldots, 1, 0$ and $l = n_L - 1, \ldots, 1, 0$.

3 Compute, from the keys obtained in step 2, the packet key k^{rlp} using (13.6), and then decrypt the packet D^{rlp} for $r = 0, 1, \ldots, r'$; $l = 0, 1, \ldots, n_L - 1$ and $p = 0, 1, \ldots, n_P - 1$.

Case 2: A user requests access to the image of resolution r' and layer l' (and with all the precincts). The key server supplies the user with keys $k^{rl'}$, $r = 0, 1, ..., r'$. The user then computes all the necessary packet keys using (13.5) and (13.6).

Case 3: A user desires access to the image of resolution r', layer l' and $m \leq n_P$ precincts P_p, $p = p_1, p_2, \ldots, p_m$. The key server replies with the keys k^{rlp} for $r = 0, 1, \ldots, r'$, $l = 0, 1, \ldots, l'$ and $p = p_1, p_2, \ldots, p_m$. The user simply uses these keys to decrypt the corresponding packets in order to obtain the desired image.

Figure 13.16 depicts an example rooted tree for a code-stream with $n_R = 3$, $n_L = 3$, and $n_P = 2$. The key associated with each node is shown in the parentheses next to the node. To access the image with resolution 1, a user only needs to know k^1; to access the image with resolution 1 and layer 1, the user needs to know k^{11} and k^{01}; to access the image with resolution 1, layer 1 and precinct 0, the user needs to know keys k^{110}, k^{100}, k^{010} and k^{000}.

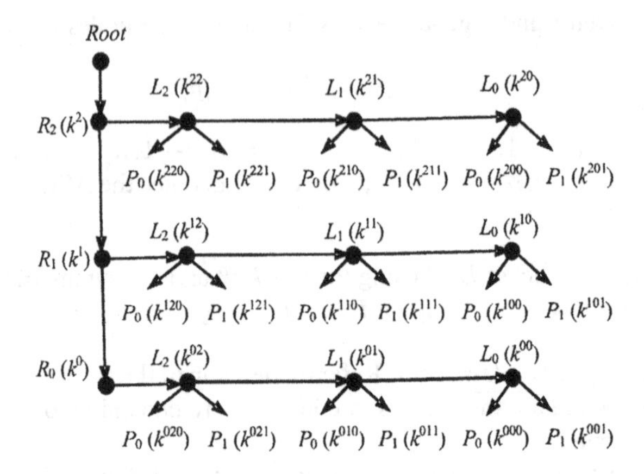

Figure 13.16. An example rooted tree for a code-stream with $n_R = 3$, $n_L = 3$, and $n_P = 2$.

5.3 Discussion

This scheme allows privileged users to access images with any resolution, layer and precinct as well as their combinations; therefore, it is very flexible and maintains the "*compress once, decompress many ways*" merit of the JPEG2000 standard. The scheme in this section is protected from the collusion attack discussed in [Wu and Deng, 2003] since its key generation process is strictly sequential and is free from the combining operation as in the second scheme.

The overhead for key transmission from the key server to a user depends on the type of images requested and on the way the rooted tree is constructed. For the tree in Figure 13.15, to access the image with resolution r', only one key $k^{r'}$ is required; to access the image with resolution r' and layer l', $(r'+1)$ keys, $k^{rl'}$, $r = 0, 1, \ldots, r'$, need to be sent to the user; however, to access the image of resolution r', layer l' and m precincts $P_p, p = p_1, p_2, \ldots, p_m$, the key server has to transmit $(r'+1)(l'+1)m$ keys, k^{rlp} for $r = 0, 1, \ldots, r'$, $l = 0, 1, \ldots, l'$, and $p = p_1, p_2, \ldots, p_m$ to the user. Therefore, the tree in Figure 13.15 is the most efficient in accommodating resolution requests and the least efficient for handling precinct requests. In general, we can easily adapt our rooted tree construction according to user request patterns. To keep the paper compact, however, we will omit the discussion here.

6. Conclusion

Based on the state-of-the-art wavelet technology, the JPEG2000 is an emerging international standard for image compression. Part 8 of the JPEG2000 standard is work in progress and is concerned with JPEG2000 code-stream security with particular emphasis on flexible authentication and access control.

The work presented in this paper is our response to the call for proposal of the JPEG2000 Security (JPSEC) Working Group. Our authentication uses Merkle tree which is optimized to accommodate the data structure of JPEG2000 code-stream. Our access control scheme uses hash functions and rooted trees for systematic key generation and packet encryption. We have implemented our schemes in a prototype which demonstrated the practical feasibility and compatibility of the proposed schemes with JPEG2000 standard Part 1.

Media streaming is becoming increasingly popular due to the explosive growth of Internet and multimedia processing technologies and applications. It would be interesting to extend our authentication and access control techniques to Motion JPEG2000 streaming in lossy networks. The technical challenge is how to make the security solutions compatible with Motion JPEG2000 functionality, efficient in processing, and more importantly, robust over lossy networks.

References

[Taubman and Marcellin, 2000] D. S. Taubman and M. W. Marcellin, (2000). , *JPEG2000 - Image Compression Fundamentals, Standards and Practice*, Kluwer Academic Publishers.

[Rabbani and Joshi, 2002] M. Rabbani and R. Joshi, (2002). An overview of the JPEG 2000 still image compression standard, *Signal Processing: Image Communication*, Vol. 17, No. 1, pages 3-48, Elsevier.

[ISO154447] ISO 154447 ITU-T Recommendation T.800, http://www.jpeg.org

[Sandhu, 1998] R. S. Sandhu, (1988). Cryptographic implementation of a tree hierarchy for access control, *Information Processing Letters*, 27(2), pages 95-98.

[Eertino et al., 1993] E. Bertino, S. Jajodia, and P. Samarati, (1993). Access Controls in Object-Oriented Database Systems – Some Approaches and Issues, in *Advanced Database Systems*, N. R. Adam and B. Bhargava, eds., Springer-Verlag Lecture Notes in Computer Science, Vol. 759, pages 17-44.

[Sandhu and Samarati, 1994] R. S. Sandhu and P. Samarati, (1994). Access control: principle and practice, *IEEE Communications Magazine*, Vol. 32, No. 9, pages 40-48.

[Akl and Taylor, 1983] S. G. Akl and P. D. Taylor, (1983). Cryptographic solution to a problem of access control in a hierarchy, *ACM Transactions on Computer Systems*, 1(3), pages 239-248.

[Chick and Tavares, 1990] G. C. Chick and S. E. Tavares, (1990). Flexible access control with master keys, In G. Brassard, editor, *Advances in Cryptology: Proceedings of Crypto'89*, LNCS 435, pages 316-322, Springer-Verlag.

[Harn and Lin, 1990] L. Harn and H. Y. Lin, (1990). A cryptographic key generation scheme for multi-level data security, *Journal of Computer and Security*, 9(6), pages 539-546.

[MacKinnon et al., 1985] S. J. MacKinnon, P. D. Taylor, H. Meijer and S. G. Akl, (1985). An optimal algorithm for assigning cryptographic keys to access control in a hierarchy, *IEEE Transactions on Computers*, C-34(9), pages 797-802.

[Ohta et al., 1991] K. Ohta, T. Okamoto and K. Koyama, (1991) Membership authentication for hierarchical multigroup using the extended Fiat-Shamir scheme, In I. B. Damgard, editor, *Advances in Cryptology: Proceedings of Eurpcrypt'90*, LNCS 473, pages 316-322, Springer-Verlag.

[Rivest et al., 1978] R. L. Rivest, A. Shamir and L. Adleman, (1978). A method for obtaining digital signatures and public-key cryptosystems, *Communications of the ACM*, Vol. 21, No. 2m pages 637-647.

[Grosbois et al., 2001] R. Grosbois, P. Gerbelot and T. Ebrahimi, (2001). Authentication and Access Control in the JPEG 2000 Compressed Domain, *Proc. of the SPIE 46th Annual Meeting, Applications of Digital Image Processing XXIV*, Vol. 4472, pages 95-104.

[Wu and Deng, 2003] Y. Wu and R. H. Deng, (2003). A method for JPEG2000 access control, *ISO/IEC JTC 29/WG1/N2810*, January 23rd.

[Prandolini et al., 2002] R. Prandolini, S. Houchin, G. Colyer (JPIP Editors), (2002). JPEG2000 image coding system - Part 9: Interactivity tools, APIs and protocols - Working Draft version 2.0, *ISO/IEC JTC 1/SC 29/WG1/N2790*, 24 October.

[Mori and Kawahara, 1990] R. Mori and M. Kawahara, (1990). Superdistribution: the concept and the architecture, *IEIEC Transactions*, Vol. E73, No. 7, July.

[Sandhu, 1993] R. S. Sandhu, (1993). Lattice-based access control models, *IEEE Computer*, Vol. 26, No. 11, pages 9-19, Nov.

[Ray and Narasimhamurthi, 2002] I. Ray, I Ray and N. Narasimhamurthi, (2002). A cryptographic solution to implement access control in a hierarchy and more, *Proceedings of the 7th ACM Symposium on Access Control Models and Technologies*, pages 65-73.

[NIST, 1995] National Institute of Standards and Technology, (1995). Secure hash standard (SHS), *FIPS Publication 180-1*, 1995.

[Rivest, 1992a] R. Rivest, (1992). The MD5 message digest algorithms, *IETF RFC 1321.*

[Merkle, 1989] R. C. Merkle, (1989). A certified digital signature, *Proc. of Advances in Cryptology-Crypto '89*, Lecture Notes on Computer Science, Vol. 0435, pages 218-238, Spriner-Verlag.

[Rivest, 1992b] R. L. Rivest, (1992). The RC4 encryption algorithm, RSA Data Security, Inc., March 12, (Properiety).

[Schneier, 1996] B. Schneier, (1996). *Applied Cryptography*, John Wiley & Sons.

[Menezes et al., 1996] A. J. Menezes, P. C. van Oorschot, and S. A. Vanstone, (1996). *Handbook of Applied Cryptography*, CRC Press.

[Lin and Chang, 2000] C. Y. Lin and S. F. Chang, (2000). Semi-Fragile Watermarking for Authenticating JPEG Visual Content, *SPIE Security and Watermarking of Multimedia Contents II EI '00.*

[Devanbu et al., 2001a] P. Devanbu, M. Gertz, A. Kwong, C. Martel, G. Nuckolls and G. Stubblebine, (2001). Flexible authentication of XML documents, *Proc. of the 8th ACM conference on Computer and Communication Security*, pages 136-145.

[Goodrich et al., 2001] M. T. Goodrich, R. Tamassia, and A. Schwerin, (2001). Implementation of an Authenticated Dictionary with Skip Lists and Commutative Hashing, *Proc. of DISCEX II'01*, Vol. 2, pages 1068-1083.

[Naor and Nissim, 1999] M. Naor and K. Nissim. (1999). Certificate Revocation and Certificate Update, *Proc. of the 7th USENIX Security Symposium*, pages 217-230.

[Devanbu et al, 2001b] P. Devanbu, M. Gertz, C. Martel and S. Stubblebine, (2001).Authentic Third-party Data Publication, *Proc. of the 14th IFIP WG11.3 Working Conference in Database Security*, IFIP Conference Proceedings, Vol. 201, pages 101-112, Kluwer.

[Peng et al., 2003] C. Peng, R.H. Deng, Y. Wu and W. Shao, (2003). A flexible and scalable authentication scheme for JPEG2000 image codestreams, to appear in *the Proceedings of the ACM Multimedia 2003*, pages 433-441, San Franciso.

[Fukuhara and Singer, 2003] T. Fukuhara and D. Singer, (2003). 15444-3 amendment 2, Motion JPEG2000, Motion JPEG2000 version 2, MJP derived from ISO media file format, *ISO/IEC JTC 1/SC 29/WG1 N2780F*, January.

[Rivest, 1992a] R. Rivest. (1992). The MD5 message-digest algorithms, RFC 1321.

[Merkle, 1989] R. C. Merkle. (1989). A certified digital signature. Proc. of [illegible] Cryptology — Crypto '89, Lecture Notes in Computer Science, vol. 435, pages 218-238. Springer-Verlag.

[Rivest, 1992b] R. L. Rivest. (1992). The RC4 encryption algorithm. RSA Data Security, Inc., March (Proprietary).

[Schneier, 1996] B. Schneier. (1996). Applied Cryptography. John Wiley & Sons.

[Menezes et al., 1996] A. J. Menezes, P. C. van Oorschot, and S. A. Vanstone. (1996). Handbook of Applied Cryptography. CRC Press.

[Tan and Chen, [illegible]] [illegible] Lodha. (2004). [illegible] in [illegible] and [illegible] NSF [illegible] and [illegible].

[illegible] [illegible] [illegible]. Matsuoka, [illegible] [illegible] [illegible]. Proc. [illegible] of SM[illegible] [illegible] [illegible] Symposium, pages [illegible].

[illegible] et al., 2001] M. [illegible] R. [illegible] and [illegible]. (2001). In [illegible] [illegible] [illegible] of [illegible] [illegible], vol. [illegible], pages 106-[illegible].

[illegible] (1999) [illegible] [illegible] pages 1-17.

[illegible] et al., 2001] [illegible] M. [illegible] and [illegible] Stubblefield. (2001). [illegible] IEEE [illegible] Proceedings, vol. [illegible], pages [illegible].

[Peng et al., 2002] [illegible] R. [illegible] and W. [illegible] [illegible] and [illegible] for IP [illegible] [illegible].

[Blackhand and [illegible], [illegible]] [illegible] and L. [illegible] [illegible] 2, [illegible] RFC 2960 [illegible] RFC 2[illegible].

Chapter 14

A SECRET INFORMATION HIDING SCHEME BASED ON SWITCHING TREE CODING

Chin-Chen Chang,[1] Tzu-Chuen Lu [1] and Yi-Long Liu [1]
[1] *Department of Computer Science and Information Engineering, National Chung Cheng University*
{ ccc, ltc, lylu91 } @cs.ccu.edu.tw

Abstract Vector quantization (VQ) is an efficient lossy image compression approach based on the principle of block coding . In a VQ system, a host image is transformed into a series of indices. In order to improve the compression rate, switching-tree coding (STC) was designed to encode the output codevector indices. In this paper, we propose a novel lossless hiding scheme. When this scheme is used, information is hidden in STC compressed codes to broaden the efficiency of the compressed codes. According to the experimental results, we find that information can be efficiently hidden in the indices without the content being modified. After decoding, the hidden information and the indices can be obtained. Only a small number of extra bits are needed to record the hidden information.

Keywords: data compression , information hiding, vector quantization, switching-tree coding

1. Introduction

A digital image is presented using a set of ordered pixels. In order to reduce the storage of pixels, many image compression methods have been proposed to eliminate redundant information of the image [Linde et al., 1980, Nasrabadi and King, 1988]. VQ is an efficient image compression method that can reduce the number of bits required to represent an image. Due to VQ's high compression ratio, VQ has been widely used in various applications, such as image and voice compression and voice recognition . In a VQ system, an input image is first divided into a set of blocks, which are called vectors, with k-dimension. Next, each vector is mapped onto a corresponding index that indicates the location of the reference vector in a codebook based on the minimum Euclidean distance criteria. Hence, the indices are used to replace all the blocks of the original pixels of the input image.

In order to compact the indices of VQ, some researchers proposed their own methods, such as switching-tree coding (STC) and search-order coding (SOC), to re-encode the indices. Hsieh and Tsai proposed the SOC algorithm in 1996 [Hsieh and Tsai, 1996]. They exploited the interblock correlation in the index domain rather than in the pixel domain. Sheu et al. proposed the STC algorithm in 1999 to re-encode the output vector indices after VQ compression [Sheu et al., 1999]. They constructed three binary trees to allocate the optimal variable-length noiseless code for each index. Both the SOC and STC algorithms utilize the high correlation between neighboring blocks to encode the VQ indices with fewer bits. In terms of the compression process of STC, we find that there are some compressed codes in the indices that are useless. If these codes can be wisely used to hide information, then the bandwidth efficiency of the compressed codes may be increased. In this paper, we propose a novel concept, that is, hiding information in the compressed codes of STC. In addition, the proposed method does not modify the contents of the information and the compressed image file.

2. Related Work

A VQ system is composed of two operations: an encoder and a decoder [Gersho and Gray, 1992, Linde et al., 1980]. The encoder maps an original image O in the vector space onto a finite set of vectors CB = $\{c_0, c_1, \ldots, c_{M-1}\}$, where CB is called a codebook, and c_i is a k-dimension vector (codeword), with $0 \leq i \leq M-1$, to form a small-sized index table. The image O is first divided into H$\times$W blocks, called vectors, where each block has h$\times$w pixels. Let $O' = \{B_{00}, B_{01}, \ldots, B_{H-1,W-1}\}$ denote the divided image, where B_{ij} is the vector in row i and column j. Then each vector B_{ij} is mapped onto a corresponding index I_{ij} by a search of the minimally distorted co-vector from CB. Let I = $\{I_{00}, I_{01}, \ldots, I_{H-1,W-1}\}$ denote the set of indices, called the index table, of the image O$'$.

Sheu et al. adopted the correlation property of adjacent blocks, which is that many neighboring blocks may be quantized into the same index to re-encode the indices of VQ [Sheu et al., 1999]. They categorized the relationships between two neighboring indices into four categories: upper connection (UC), left connection (LC), around connection (AC), and disconnection (DC).

Let P be the input index value of I_{xy}. Let L, which is to the left of P, be the neighboring index value of $I_{x-1,y}$. Let U, which is above P, be the neighboring index value of $I_{x,y+1}$. And let A, which is the same as P, be the around index value of $I_{x+k,y+k}$. The rules used to determine the relationship between two neighboring indices are given below: If U=P, then the index I_{xy} and the index

$I_{x,y+1}$ has the relationship UC. If L=P, then the index I_{xy} and the index $I_{x-1,y}$ has the relationship LC. If A=P, then the index I_{xy} and the index $I_{x+k,y+k}$ has the relationship AC. Otherwise, the index I_{xy} has the relationship DC with neighboring indices.

For example, assume that an index table I of size 8×8, as shown in Fig. 14.2, where P is the index value (I_{44}), U is the upper index value of P (I_{34}), and L is the left index value of P (I_{43}). If P = 7, which is the same as U, then P and U have the relationship UC. If P = 10, which is the same as L, then P and L have the relationship LC. If P = 18, which is the same as A in the around indices (I_{12}) of P, then P and A have the relationship AC. However, if P = 50, which is completely different from all indices in the neighboring area, then P and it's neighboring indices have the relationship DC.

Based on these four relationships, Sheu et al. constructed three binary connection trees to encode the indices. The trees are shown in Fig. 14.1. In the figure, A is the around index of P. In order to check whether the nearby area of P has the same value as P, Sheu et al. predefined a search path, which is shown in Fig. 14.2. In the figure, (x) means the index can be excluded, since the same index value is in the previous indices. The numbers in the range (0)~(7) indicate the corresponding addresses of P. In Fig. 14.1 (a), if P and U (or L) has the relationship UC (or LC), then P is encoded as '11' (or '10'). For example, if P = 7 in Fig. 14.2, then the compressed code of P is '11'. If P and A has the relationship AC, then the compressed code of P is composed of the prefix codes '01' and the index address of A in the binary system. For example, if P = 14, according to the predefined search path, we find that the index value of I_{33} is the same as P, where I_{33} is in the 3rd corresponding address of P. Assume that the previous indices that can be searched in the nearby area are 32 bits. The number of bits to represent the address of this index is 5 bits, since $\log_2 32 = 5$. Then the compressed code of P is '01'||(3) = '0100011.'

In this paper, we propose an efficient hiding scheme that is based on the STC compression method. The proposed scheme quickly compresses the index table of VQ into small-sized compressed codes and effectively hides information in the compressed codes.

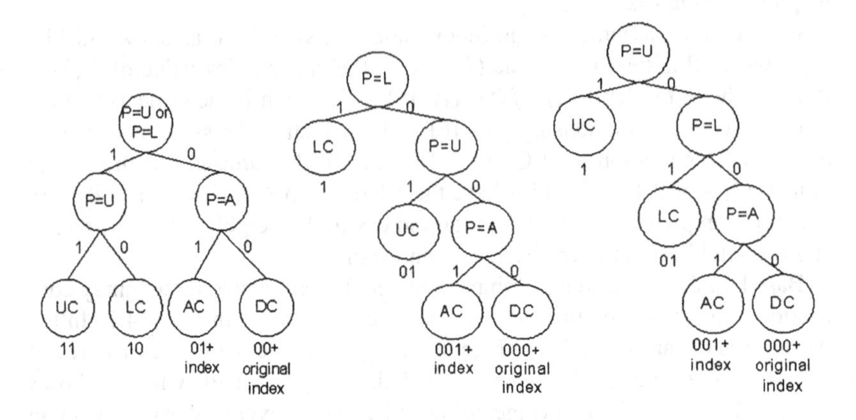

Figure 14.1. Three binary connection trees.

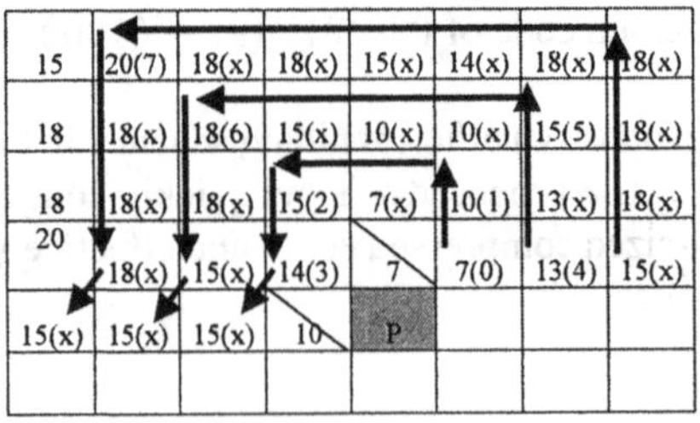

15	20(7)	18(x)	18(x)	15(x)	14(x)	18(x)	18(x)
18	18(x)	18(6)	15(x)	10(x)	10(x)	15(5)	18(x)
18	18(x)	18(x)	15(2)	7(x)	10(1)	13(x)	18(x)
20	18(x)	15(x)	14(3)	7	7(0)	13(4)	15(x)
15(x)	15(x)	15(x)	10	P			

Figure 14.2. The search order of STC.

P: The current index ⟶ : Search path; (0)~(7): The corresponding address of P; (x): The exclusion of repetition searched indices

The Proposed Method

In order to hide information in the compressed codes of STC, we propose a novel information-hiding scheme, which is called IHSTC. Assume that a sender wants to send an original image and secret information to a receiver. Let O denote the original image and S = $s_1s_2 \ldots s_K$ be the secret information, where $s_j \in [0, 1]$, 1≤j≤K, and the length of S is K. In which S is obtained by encrypting a plaintext T = $\{m_1m_2 \ldots m_K\}$ using a DES-like method associated with the private key, where $m_j \in [0, 1]$ and 1≤j≤K. In order to save the time that is required in transmitting, the image O is compressed into small-sized indices by VQ compression method. Let I = $\{I_{00}, I_{01}, \ldots, I_{H-1,W-1}\}$ be a set of the indices, where I_{ij} ∈[0, M-1], M is the size of the codebook, H is the height of I and W is the width of I. After VQ compressing, the image O is transformed into the index table I. Then IHSTC encoder compacts the index table I into small-sized compressed codes and hides S in those codes at the same time.

2.1 Encoder of IHSTC

In this subsection, we shall describe how to re-encode I into small-sized compressed codes and hide information S in it at the same time. We only use Tree A to re-code the indices. The encoder uses three binary connection trees that can follow the same way. In Tree A, there are four cases to encode an input index.

Case 1: P has relation UC to U, where P=U, then P′='11'.
Case 2: P has relation LC to L, where P=L, then P′='10'.
Case 3: P has relation AC to A, where P=A, then P′='01'||β(γ(P)).
Case 4: P has relation to other indices, then P′='00'||β(P).

The function γ returns the address of A in the neighboring area of P except U and L. If the same value as P cannot be found, then the function γ returns '-1'. The function β(x) returns the binary string of the value x. If P has relation UC, LC or AC to other index in the neighboring area, the compressed code of P can be used to hide information s_k. The compressed code after hiding is called P′. Otherwise, the compressed code of P cannot hide any information and P′ is formed by concatenating by '00' with its original index in the binary system. In addition, if the secret value of information s_k is '0', then P is compressed normally. Or P needs to give up the first equal index and to hide s_k in the second equal index. IF P can find more than two equal indices, then P hides s_k in the second index rather than the first index. For example, assume P in Fig. 14.2 is 7. If s_k = '0', then P is compressed normally by Case 1, since P = U, and P′ = '11'. However, if s_k = '1', then P gives up the first index U and searches next equal index in the neighboring area. The second index exists in the zeroth corresponding address of P (I_{35}). Therefore, P′ = ‘01’ $\|\beta$ (0) = ‘0100000’.

2.2 Decoder of IHSTC

When the receiver receives FC, the decoder is needed to decompress the original index table and to extract the hidden information. The process of decompressing and extracting is similar to that of compressing and hiding by using IHSTC. Since there are only four different encoding cases, the indicators are always presented in 2 bits. In FC, the first two bits, FC_0 and FC_1, are for the first indicator that is used to determine the length of the first index. The length, t, should be 8, C, or 0. The bit sequence, $FC_2FC_3KFC_{t+1}$, is then transformed to be a decimal number as the first index. The following two bits, FC_{t+2} and FC_{t+3}, are then for the next index. Via performing this process continuously, the indices can be decoded correctly from the compressed code FC. In other words, in the decoder, FC_iFC_{i+1} indicates P is encoded by which case in Tree A. For example, if FC_iFC_{i+1} equals ‘00’, then P is obtained from FC_{i+2} to FC_{i+9}. If we can find another index, which has the same value as P in the neighboring area (γ(P)$\geq$0), it means that the hidden information is ‘1’ (s_k = 1). Then let the pointer i point to next indicator, where i = i + 10. Similarly, if FC_iFC_{i+1} equals ‘01’, P is obtained from FC_{i+2} to FC_{i+C+1}; if FC_iFC_{i+1} equals ‘10’, P is obtained from the left index L; if FC_iFC_{i+1} equals ‘11’, P is obtained from the upper index U.

3. Experiments

A system for compressing and hiding index tables, based on IHSTC and called the IHSTC system, was developed on a PIII 450 MHz personal computer with 128 Ram and was tested on several image files. Before the results of the system are described, several notations are first defined below.

3.1 Notation Definition

Let S be the secret information and $|S|$ be its length. PSNR is the peak signal to noise ratio (PSNR) value of the image after VQ compressing. Let the symbol ‘C’ be the number of bits for representing the size of the neighboring area of P. For example, assume the neighboring area of P is 32, then C = 5 (2^5 = 32). NSTC is the total number of indices, which has relation UC, LC or AC to other index, in an index table. It also equals the maximum length of the secret information that can be hidden in the index table. Several image files were used for compressing and hiding to evaluate the performance of the IHSTC system. The images are shown in Fig. 14.3.

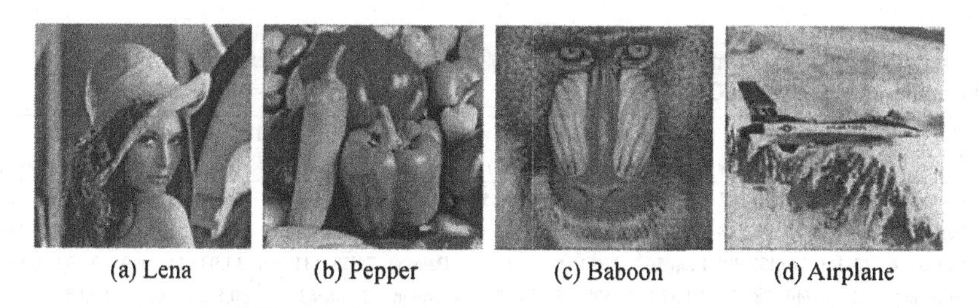

(a) Lena (b) Pepper (c) Baboon (d) Airplane

Figure 14.3. The experimental images.

3.2 The Experimental Results

In order to find the most proper C value, different bit lengths of 4 and 5 were set to run the different sizes of image files in the IHSTC system. In the first experiment, 2048 bits are hided in the index table of the image files, with 512×512 pixels, where $|S| = 2048$. The experimental results are shown in Tables 14.1, and 14.2, in which $C's$ in Table 14.1 and Table 14.2 are 4 and 5, respectively. The column $|STC|$ represents the total number of bits of the index table after STC compressing, and the column $|IHSTC|$ represents the total number of bits of the index table after IHSTC compressing and hiding. The column δ is the difference of bits between the STC compressed codes and the IHSTC compressed codes, which is equal to $|IHSTC|$ - $|STC|$. The sub-column 'Tree A' represents the index table compressed only by Tree A. On the other hand, '3 Trees' represents the index table compressed by three binary connection trees.

The information load of an image is the maximum length of the information string, which can be hidden in the image. The load in the IHSTC system is based on NSTC. The third experiment was performed to calculate the required extra bits of the images under the maximum information load. Tables 14.3, and 14.4 show the experimental results, where $|S|$ is equal to the NSTC of each image.

Table 14.1. The experimental results of the images with $|S|$=2048 and C=4.

Image	NSTC	|STC|		|IHSTC|		δ	
		Tree A	3 Trees	Tree A	3 Trees	Tree A	3 Trees
Lena	12,132	86,280	86,202	89,516	90,125	3,236	3,923
Pepper	11,451	90,076	90,528	93,048	93,813	2,972	3,285
Baboon	6,527	130,376	132,328	133,580	135,444	3,204	3,116
Airplane	12,111	80,740	78,209	83,072	81,216	2,332	3,007

Table 14.2. The experimental results of the images with $|S|$=2048 and C=5.

Image	NSTC	|STC|		|IHSTC|		δ	
		Tree A	3 Trees	Tree A	3 Trees	Tree A	3 Trees
Lena	12,352	90,494	90,416	92,222	92,555	1,728	2,139
Pepper	11,685	93,835	94,287	95,498	96,074	1,663	1,787
Baboon	7,553	131,986	133,938	133,270	135,195	1,284	1,257
Airplane	12,388	83,356	80,825	84,907	82,574	1,551	1,749

Table 14.3. The experimental results of the images with $|S|$=NSTC and C=4.

Image	NSTC	|STC|		|IHSTC|		δ	
		Tree A	3 Trees	Tree A	3 Trees	Tree A	3 Trees
Lena	12,132	86,280	86,202	102,004	104,746	15,724	18,544
Pepper	11,451	90,076	90,528	105,672	108,067	15,596	17,539
Baboon	6,527	130,376	132,328	140,856	142,351	10,480	10,023
Airplane	12,111	80,740	78,209	94,528	95,572	13,788	17,363

Table 14.4. The experimental results of the images with $|S|$=NSTC and C=5.

Image	NSTC	|STC|		|IHSTC|		δ	
		Tree A	3 Trees	Tree A	3 Trees	Tree A	3 Trees
Lena	12,352	90,494	90,416	107,340	109,918	16,846	19,502
Pepper	11,685	93,835	94,287	110,283	112,807	16,448	18,520
Baboon	7,553	131,986	133,938	141,889	143,393	9,903	9,455
Airplane	12,388	83,356	80,825	99,051	100,087	15,695	19,262

Some experimental observations are given below. 1) The NSTC of an image is directly proportional to the image size. The information load in a large image is greater than that in a small image. 2) Increasing the number of C will increase the number of NSTC. That means increasing the number of C can extend the information load of images. Nevertheless, extra bits may be required to represent an index. 3) The IHSTC system obtains better results when C = 4. 4) The IHSTC system using Tree A leads to better results than using three binary connection trees. 5) The PSNR value of the image Baboon is lower than that of the other images. It is the only image for which the IHSTC system gets better results when three binary trees are used. 6) The IHSTC system can hide a huge amount of information in the index table of an image file, and only a few extra bits are needed to record the corresponding information.

4. Conclusions

In this paper, we proposed a novel information-hiding scheme that is based on a switching-tree coding, called IHSTC. IHSTC is able to completely recover the index table of image files, which is compressed by VQ, and to hide information in it. The average time needed to hide an information character in an index table by using the IHSTC system is 0.077 seconds. From the experimental results, it is obvious that IHSTC is indeed an efficient and effective scheme for hiding secret information in image files.

References

[Gersho and Gray, 1992] Gersho, A., and Gray, R.M. (1992). *Vector Quantization and Signal Compression*, Kluwer Academic Publishers.

[Hsieh and Tsai, 1996] Hsieh, C. H. and Tsai, J. C. (1996). Lossless Compression of VQ Index with Search-order Coding. *IEEE Transactions on Image Processing*, Vol. 5, No. 11, pages 1579-1582.

[Linde et al., 1980] Linde, Y., Buzo, A. and Gray, R. M. (1980). An Algorithm for Vector Quantizer Design. *IEEE Transactions on Communications*, Vol.

28, No. 1, pages 84-95.

[Nasrabadi and King, 1988] Nasrabadi, N. M. and King, R. B. (1988). Image Coding Using Vector Quantization: A Review. *IEEE Transactions on Communications*, Vol. 36, No. 8, pages 957-971.

[Sheu et al., 1999] Sheu, M. H, Tsai, S. C. and Shieh, M. D. (1999). A Lossless Index Coding Algorithm and VLSI Design for Vector Quantization. *Proceedings of the 10th VLSI/CAD Symposium*, pages 485-488.

Index